Trans*cognitive© *Spirituality

Shaping the New Paradigm at the Interface of Consciousness and Reality

To Mhaletta,
with love & blessings.
Robert x

Robert Goodwin

Transcognitive Spirituality

ISBN 978-0-9572745-1-8

Cover montage by Robert Goodwin, incorporating images downloaded with permission from iStockphoto.com

Other books by the same author:

Truth from the White Brotherhood (1998)
The Golden Thread (1999, reprinted 2005)

With Co-author Amanda Goodwin

Answers for an Enquiring Mind (2002)
In the Presence of White Feather (2005, reprinted 2010)
The Enlightened Soul (2008)
The Collected Wisdom of White Feather (2010)

Published by R.A. Associates 2013
Designed and printed in the UK
mail@whitefeather.org.uk

"Science is not only compatible with spirituality; it is a profound source *of* spirituality."

- Carl Sagan (1934-1996)

science ***noun*** **1** the systematic observation and classification of natural phenomena in order to learn about them and bring them under general principles and laws. **2** a department or branch of such knowledge or study developed in this way, eg astronomy, genetics, chemistry. **3** any area of knowledge obtained using, or arranged according to, formal principles • political science. **4** acquired skill or technique, as opposed to natural ability. **sciential** ***adj*** referring or relating to science; scientific.
ETYMOLOGY: 14c: from Latin scientia knowledge, from scire to know.

spiritual ***adj*** **1** belonging, referring or relating to the spirit or soul rather than to the body or to physical things. **2** belonging, referring or relating to religion; sacred, holy or divine • a spiritual leader. **3** a belonging, referring or relating to, or arising from, the mind or intellect; b highly refined in thought, feelings, etc. **4** belonging, referring or relating to spirits, ghosts, etc • the spiritual world. noun (also Negro spiritual) a type of religious song that is characterized by voice harmonies and which developed from the communal singing traditions of Black people in the southern states of the USA. **spirituality** ***noun.*** **spiritually** ***adverb.*** **spiritualness noun.**
ETYMOLOGY: 13c: from Latin spiritualis; see spirit.

converge ***verb*** **(converged, converging)** intrans **1** (often **converge on** or **upon someone** or **something**) to move towards or meet at one point. **2** said eg of opinions: to tend towards one another; to coincide. **3** said of plants and animals: to undergo convergent evolution. **convergence** ***noun.***
ETYMOLOGY: 17c: from Latin convergere to incline together.

Definitions taken from The Chambers Dictionary (online version) http://www.chambers.co.uk/dictionaries/the-chambers-dictionary.php

This book is dedicated to my close family, with my boundless love and gratitude.

With grateful thanks to my dear friend Jos.

"We are not born into the world. We are born into something that we make into the world."

- Michael Talbot (1953-1992)

Contents

Overview

If you're anything like me, you don't like reading the 'Introduction' to a book and would probably prefer to get to the meaty stuff straightaway. Well, bear with me because I'll keep this overview as brief as possible as I set the scene for what is to follow and you'll find that things will move along quite quickly.

Anyone, having demonstrated publically as a medium for over thirty-five years would be certain to have tales to tell and wonders to reveal about their connections with the afterlife and the many extraordinary events they had experienced during that time. They might even be forgiven for wanting to write their memoirs or share stories with those wishing read about their exploits. Let's face it, the majority of people like to talk about themselves and what they've accomplished over the course of their lives, naturally believing that others will be eager to know too. One only has to visit any high street book store to witness shelves lined with autobiographies from sports stars, showbiz celebrities, cooks, gardeners, politicians and yes, psychics and mediums. Fortunately you can heave a sigh of relief because I have no such desire and you will be pleased to know that this book is not about me.

Having cleared that up we can both relax a little as we begin this exploration into what for some, will be uncharted territory and for others, may be an extension of what they have already begun to discover for themselves. Quite where this journey will lead us, I'm not quite certain because if I have learned anything at all in my soon to be sixty-years upon this planet (and perhaps many more before I came here this time around) it is that one should never pre-judge anything or set any limits on where consciousness can go. To borrow a phrase from my days as an NLP practitioner; there may be some new 'learnings' that await us as we turn the pages of this book together.

Firstly though, the title; I've not set out to be deliberately clever or to confuse the reader in any way. I'd simply like to think the words themselves emphasise the deeper meaning underpinning the themes that are contained throughout. In simple terms, *Trans* means to 'cross over' or to 'be across something' whilst *cognitive* is connected to thinking and conscious mental processes. *Spirituality* of course, covers a broad spectrum of meaning which according to a certain popular encyclopedia refers to 'an ultimate or alleged immaterial reality or inner path, enabling a person to discover the essence of their higher being' or the 'deepest values and meanings by which people live.' Sounds about right to me. Hence, *Transcognitive Spirituality* encompasses the connection between and across our inner states of consciousness and the relationship they have with the journey that we are all engaged upon.

For many years mainstream science seemed to suggest that there was no direct correlation between what was perceived as the 'outer' reality and the 'inner path', or to put it more bluntly, no connection between what appeared external to the human condition and that which lay within. The very universe itself, its origin having been defined by the widely accepted *Big Bang* theory of creation, seemed for many to be the macrocosm out of which life accidentally arose and in which it merely existed until being extinguished at death. We could not it seemed, influence or interact with it to any great extent. Even God, whose existence remains scientifically unproven until this day had been confined to the realms of religious belief whilst more popular theories such as Darwinism became increasingly prevalent and more widely accepted.

There has though, begun a shift in awareness that is bringing about enormous changes in human understanding as new and wonderful revelations unfold before us, for not only are we connected to everything that is, *we are everything that is.* However, many challenges lie ahead if we are to fully understand and integrate this information and it will inevitably lead to major upheavals and enormous changes in the way that we view 'reality'.

For my part, there have been many such shifts that have lead me from one level of awareness to another and I have always felt a sense of being directed to information that it was necessary for me to know at any given moment. Thus, my own views on the nature of existence have been shaped not only by my observations of life but more importantly by my inner, subjective experiences and guidance. Some of the themes and concepts I will be sharing with you cannot yet be supported by scientific fact, but are nonetheless worthy of consideration and deeper scrutiny and I would urge you to evoke your own innate sense of knowing to discern whether or not you feel that my ideas are valid and whether they simply 'feel' right.

I will attempt to combine the scientific and the esoteric in as seamless a way as possible, especially as I believe that because they are parts of the whole, each compliments the other. You will notice that I have also used illustrations throughout to give emphasis and clarity to some of the more complex points under discussion. These are by no means definitive and should be regarded more as signposts or working models rather than complete in themselves. In addition, I have borrowed some ideas from other writers and thinkers that have inspired me, developing and expanding upon their themes whilst hopefully adding something of value to them. It is hoped that the expansion of some of these concepts will lead to a deepening of your own understanding and to this end I would urge you to give consideration to delving further by whatever means inspire you.

This book is also written from the perspective of the layman. I do not have an academic background and my scientific knowledge does not ascend to the heights of a Hawking or a Dawkins. What I do have though is insight and the advantage of being predominantly right-brained. My intuitive abilities remain as sharp as they have ever been and these, when coupled with my thirst for truth and my natural instinct to reject mainstream views that do not satisfy my reasoning mind allow me to venture into areas where some would fear to tread.

A word of warning here; if you are an atheist or complete skeptic in your beliefs about life after death and cannot be appeased by more comforting terms such as 'non-physical dimensions' or 'alternate realities' then perhaps we may need to go our separate ways with no hard feelings until such time that our pathways cross once more. I won't be offended honestly, I am used to it.

One final plea then before we dive headlong into this metaphysical ocean of ideas and concepts that I will attempt to explain in ways that will excite rather than exasperate; keep your mind and most importantly your heart, open. Don't follow behind me on this journey - walk with me. If you find yourself leaping in front, go right on ahead because your consciousness may take you to places that I haven't yet dreamed of. Hopefully we may arrive together at some future point of understanding and congratulate each other at having made it. For now, grab your swim gear, take a deep breath and let's go!

Prologue: Imagine

Sitting alone, huddled against the biting wind, he scans the night sky, watching the array of twinkling stars that lie scattered across the heavens. Leaning his head to one side he peers intently into the darkness, his rudimentary thought processes struggling to comprehend the nature of the myriad of tiny specks of light that interrupt his vision. As the clouds part, revealing yet more of the vast swathe of stars that form the milky way, his primitive Neolithic brain strives to make sense of this mysterious vision, but without any relative experiences from which to draw comparisons, any attempts to analyse this panoramic vista prove futile. Yet, closing his eyes, his brain cells fire, releasing fragmented images that swirl within his consciousness and he wonders….

Over one hundred thousand years previously, in a distant galaxy a giant sun, starved of the fuel that is its life blood, collapses inward, becoming a neutron star, the resulting supernova explosion expelling vast quantities of interstellar material across space. Moving at tremendous speeds the remnants of this once gloriously illuminated body are scattered in all directions throughout the cosmos. Through the apparent emptiness of space they shoot, some no larger than a marble, others up to a mile or so across, virtually unhindered, traversing vast distances beyond other suns and planetary systems. Entering the atmosphere of planets such as our own they once more become illumined as their surfaces begin to melt under the intense pressure and they light up the sky as their course propels them ever onward to some unknown destiny.

Time elapses, with months, years and centuries passing by. Still travelling at a velocity approaching one per cent the speed of light the meteor shower eventually reaches our earth's upper atmosphere with the resulting display of shimmering lights that appear to fall from the sky like brightly coloured stones scattered from some unseen hand impacting upon the retina of his Neolithic eyes which open widely with a start. Gazing upward in awe, they take in the light show that is now expanding across the blackness of the night sky, transfixed by the brightly coloured shower that is unlike anything they have witnessed before. Trying once more to comprehend the nature of this phenomenon the primitive brain of the ancient hominid strives vainly to understand what his eyes are witnessing. In his short, undistinguished existence he has never seen anything like this before and he feels a mixture of conflicting emotions and thoughts, everything from fear to excitement as he attempts to come to terms with this cosmological event.

Then a thought, vague at first, almost intangible yet growing ever stronger, bubbles up from the recesses of his subconscious mind before fading again into obscurity…

"Is there something greater and more powerful than I?"

Much later in the history of mankind, others would have similar thoughts and the primitive concept of Gods and all that accompanied them would arise within the collective human psyche giving birth to ritualistic behaviours, human sacrifices and eventually, numerous religions.

Yet for now, this was a truly original thought. No other life form upon Earth had thought it before. A critical threshold had been reached that would see a deepening in consciousness and an evolutionary leap for the species known as humanity. As man began to evolve physically, his body changing in form and complexity to meet the demands of the environment and the needs of survival, so would there also be a deepening of his mental, emotional and spiritual characteristics that would raise him above the level of the animal from which he had emerged, transforming him into a thinking, reasoning being.

This is an eternal process. Original thought lies within the capability of us all and goes beyond the influences of conditioning brought about through our life experiences and the environment in which we exist. We may however, upon reflection, begin to glimpse something greater than thought. Something *beyond* thought, a dimension in which thought as we may comprehend it does not exist.

Its name is *infinite consciousness.*

Chapter One

The 'Reality' Inside Your Head

"Reality is merely an illusion, albeit a persistent one"

Albert Einstein (1879 – 1955)

René Descartes (1596–1650) the famous French philosopher, mathematician and writer once said 'If you would be a real seeker after truth, it is necessary that at least once in your life you doubt, as far as possible, all things.' That is quite a statement to make, because to doubt everything you have come to know and understand requires great courage and few of us are ready to take such a leap of faith. We like everything to be neatly packaged and accessible, much like the food we consume and the products we buy. Take away the familiar and what are you left with? – the unfamiliar. And who likes that?

So let's begin by familiarising ourselves with an unfamiliar landscape. Let's put away the route planner and discard the sat-nav because they won't operate where we're heading. Try this for size; the world that you see around you – the rocks, the plants, the oceans, the trees, the buildings, the animals, the birds and even the humans are not 'out there'. They are 'in here', in your head. More accurately, they are in your mind as constructs of the brain, with the real 'you', *consciousness,* acting as the observer - because we as individuals never experience any tangible reality except that which we know through consciousness.

The notion that we never directly experience 'the world' has fascinated many notable philosophers across the years. Emmanuel Kant (1724-1804), the German philosopher who lived in the eighteenth century wrote about the *phenomenon* (forms that appear in the mind) and the *noumenon* (that which is apprehended). Kant stated that the former is all that we know with the latter being always beyond our knowing. An earlier philosopher, John Locke (1632-1704) held the view that all knowledge was based upon external objects acting upon our senses but whereas he asserted that perception was passive, Kant believed that the mind was an ardent participant, actively forming the world around us. He viewed reality as something that we each construct for ourselves. For Kant to reach such conclusions was quite remarkable because he did not have the same access to scientific research or understanding of perception that we do today. Our current knowledge gives us a much clearer picture of how the brain constructs its image of reality. The brain sits in a darkened cavity known as the skull and all it ever sees are electrical and chemical signals being transmitted through the senses from which it then constructs 'reality'. This may be difficult to comprehend for many because we are so used to thinking that what we are actually interacting with when we wake up each morning is a world that is external to us, but the truth is that we are actually seeing *our own unique internal model of the world.* Even when we are asleep, our brain is still constructing images for us because it doesn't need our eyes to see and despite the absence of any visual stimuli we can still have vivid experiences whilst the same processes are running within the visual cortex. When we have dreams we are aware of sights, sounds and other sensations as well as feeling a whole range of emotions and are even able to interact with others who appear to us as separate individuals. All of this seems to be 'outside' of us and also very real to the extent that changes in the physiology of the body occur as a direct response to our experiences. Only when we awaken from the sleep state do we realise that it was all just a creation of the mind.

In the normal waking state light reflected from an object forms an image upon the retina of the eye. Photo-sensitive cells within the retina then discharge electrons which in turn trigger electro-chemical impulses that travel along the optic nerve to the visual cortex before undergoing various procedures that process the data into shapes, movements, patterns, colours and so on. The brain then integrates this information, bringing coherence as it creates a reconstruction of the 'external' world. Information passing into the brain through the other sense organs is processed in a similar fashion as the raw data is analysed and filtered to add other perceptual layers to the experience. For example; if we hold a banana in our hand, receptors in the nose will be triggered by molecules being emitted from the fruit giving us the sensation of being able to smell it, whilst other cells in our skin are simultaneously sending messages informing us of the texture, warmth, smoothness and so on, giving us an overall impression of what it is we are interacting with. In addition the brain also processes information from other parts of the body and acts as a kind of central computer that constantly monitors and regulates bodily functions, but for now we'll take this as self-evident whilst continuing our focus on how we perceive reality.

We must mention of course that other forms of information are also received by the mind and relayed to the brain from *non-physical* dimensions, although more about this later in the book. I will also introduce at this early stage the idea of a clear distinction between the brain, the mind and consciousness. These terms are so often portrayed as being one and the same and mainstream science in particular holds the collective view that the latter are both products of the former. It is my contention however that *individual consciousness,* which is what 'we' are, is part of *infinite consciousness* (God) and utilises mind (*organizing intelligence*) to function through the brain and biological computer (*human form*) at this (*physical*) level of reality. To break it down even further we could state that it is never the totality of our individual consciousness that is present within the localized form of the human body, but only a facet of it. A good analogy for this would be to consider *infinite consciousness* (God) as the ocean, *individualised consciousness* as a cup drawn from the ocean and *individual personality* (the facet expressing itself within the physical level) as a droplet. All three would be seen as both separate, yet connected to each other *(Fig 1)*

Fig 1. Three aspects of consciousness

The illusion that we perceive the physical universe directly is for the most part, entirely convincing. From time to time however, we do encounter phenomena that suggest otherwise. These can be considered as pointers, revealing other alternatives to our construction of reality. Upon initially viewing the illustration below *(Fig 2.)* most people see a white vase against a black background. Equally you may then view it as two faces, side on against a white background. Both views are true of course, but the brain usually sees the version that it is most used to first. The second image *(Fig 3.)* shows a series of lines representing a box. The question is, do you see it from above or below? If you are in the majority you will be seeing the box as if viewed from above but a change of emphasis will easily allow your brain to view it as if from beneath - either way, what is essentially a two-dimensional image is being viewed as a three-dimensional one by the brain.

Fig 2.

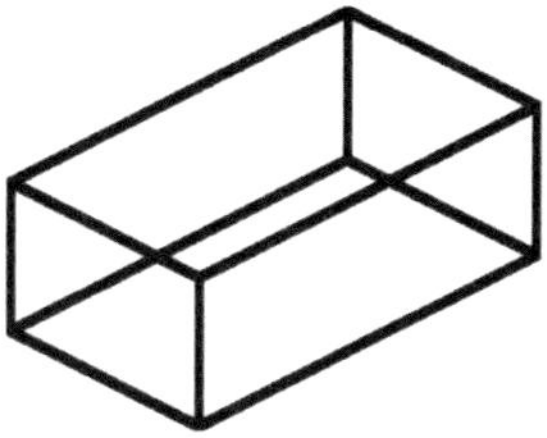

Fig 3.

There are numerous examples of this phenomena occurring and the brain will often substitute an image for one that isn't there or even ignore something that it *is* there *(Fig 4 overleaf)*.

Fig 4.

What conclusions can be drawn from this? We can assume that the cortex of the brain is very good at extracting information that is useful and has some correlation with things in the 'outside' world. This is an ability that is the result of millions of years of evolution and enables us to make more sense of this level of being. There are though, some who claim that this inner personal reality is not the same as that which is actually stimulating our senses. Shankara, the famous Hindu philosopher wrote that 'All things - from Brahma the creator down to a single blade of grass, are simply appearances and not real', whilst Niels Bohr (1885-1962), the pioneer of twentieth century physics said 'An independent reality, in the ordinary physical sense, can neither be ascribed to the phenomena nor to the agencies of observation'. It can even be argued that what we actually do see is never a direct in-the-moment event but something that has already occurred within linear time. There is after all, a miniscule time delay between perception and construction, rather like the time it takes for a live broadcast to become visible on our television screen. If this is true then we are always experiencing 'the past' and can never know 'the present' directly.

Despite this, our senses for the most part, confirm to us that there really is some kind of shared human experience taking place and there appears to be a common worldview of what is 'happening' and 'the way things are' at any given moment. This consensus exists on numerous levels, some of which we are consciously aware and others that lie beneath the surface in the realms of unconscious activity. We could term our conscious understanding 'the world view' implying that the majority of individuals share a common perspective – for example; an agreement that the earth is a sphere, as opposed to it once being thought of as flat. There is however a deeper mutual view existing below the conscious level that I refer to as *'the collective unconscious'* [1] and I propose that it is from this level that true change is instigated when a sufficient percentage of the human race at any given moment share and accept as true the same collective information. It has been estimated that only a relatively small percentage of the totality of humanity need to change their belief or become aware of new information for the entire remainder to eventually follow suit and it is as if this enlightenment is passed on unconsciously and effortlessly at the quantum level of being.

The famous hundredth monkey syndrome, originated by Lawrence Blair and Lyall Watson (1939-2008) in the late nineteen-seventies which was based on the study of Macaque monkeys by Japanese scientists went some way in attempting to explain the phenomenon by claiming that learned behaviour by a few within the group instantly spread across the remaining number to monkeys on neighbouring islands. I suggest that in the case of the human collective unconscious, information need not be restricted just to learned behaviour, but also encompasses information received from non-physical dimensions.

1. This should not be confused with Carl Jung's concept of the same name.

Many of us will be able to recall occasions when we have suddenly become consciously aware of something as if it had just popped into our mind from nowhere. If we liken the conscious mind to the surface of a lake, occasionally choppy and at other times calm and smooth and then consider information reaching it like bubbles rising from the depths and suddenly popping as they reach the top, then we can begin to see how this operates. Until the bubble of information or 'knowing' reaches the surface, we have little or no awareness of its existence. It is however, very real and like an air bubble emerging from a source deep within the ocean of our unconscious being, it cannot be contained, always attempting to escape to the light of day *(Fig 5).*

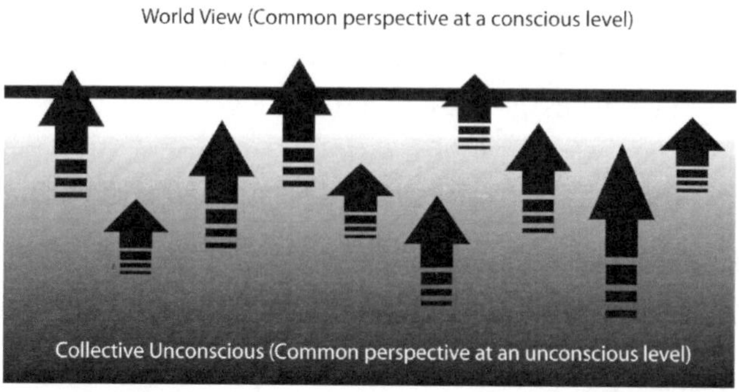

Fig 5. The world view

There are inevitably, constraints upon what we term 'our experience'. It is highly unlikely that even if enough of us wholeheartedly believed that we could walk upon water, we could actually do so. Even if we tried, there would be predictable consequences and in all probability we would end up getting very wet. The same eventuality would also undoubtedly occur if we thought we could fly and attempted to jump off a tall building and we would probably spend a long time reflecting upon the wisdom of our decision.

What is evident is that there are laws operating in every sphere of existence. What we know as *natural law* encompasses and impacts at both physical and non-physical levels. Intrinsic to all aspects of natural law is balance or as I prefer to call it, an *implicit order.* We may not always appreciate this when we are witnessing the almost mundane operation of physical laws such as when we clumsily drop a precious ornament and it smashes to pieces having been acted upon by the force of gravity or we are late for an appointment because the sun inconveniently rose at a precise time before we were ready to get out of bed. But for these laws to even exist suggests again that there is something that they need to act upon, some kind of actual tangible reality, something real.

To further complicate matters, science has delved into the heart of the quantum world only to reveal that the building blocks of substance are in themselves insubstantial. The atom is largely empty. Despite the discovery of smaller sub-atomic particles and the knowledge that orbiting electrons surround the central nucleus, there is essentially a vast inner universe of nothingness within each atom. Furthermore even the sub-atomic particles themselves are far from solid. They cannot be easily measured and they appear more as waves than particles with no definite location. It is also known that some particles appear to make spontaneous transitions between what is termed matter and anti-matter suggesting that they blink in and out of existence.

The wave/particle duality is not as confusing as it may first appear though when you begin to understand the way in which consciousness interacts at the quantum level. Viewed from an alternative perspective we may begin to understand that consciousness is integral to all of creation and that the very act of observing waveform function causes it to collapse into a particle form, or what we term 'reality'. It is worthwhile briefly mentioning here *The Copenhagen Interpretation,* - a theory that originated from discussions between Niels Bohr, Werner Heisenberg (1901-1976) and others in the latter part of 1926 and early 1927. This is a theory that helps define the science of quantum mechanics and although it is not without its flaws, remains the most popular explanation of how things work at the quantum level. We have to be mindful however that any investigation into the quantum realm will incorporate probability rather than absolute certainty. The four cornerstones of The Copenhagen Interpretation are: wave/particle duality, the probability wave function, the uncertainty principle and the importance of the observer. The theory is that the wave component of wave/particles is *probability* and the particle component is a *tendency to exist.* The famous two-slit experiment *(Fig 6.)* is essentially the best example of this.

According to the Copenhagen Interpretation the sub-atomic systems are made of entities referred to as wave/particles. These entities have both wave and particle characteristics depending upon the experiment. The illustration shows a beam of light passing through a single slit and then through two further slits resulting in an interference pattern. The resulting image on the screen reveals the characteristic lined pattern resulting from the collapse of the waveform into particle form. The fact that there is an observer interacting with this experiment is fundamental to the resulting phenomena. Interestingly however is that the main issue left unaddressed by The Copenhagen Interpretation is the causality of the wave function collapse and when it actually occurs.

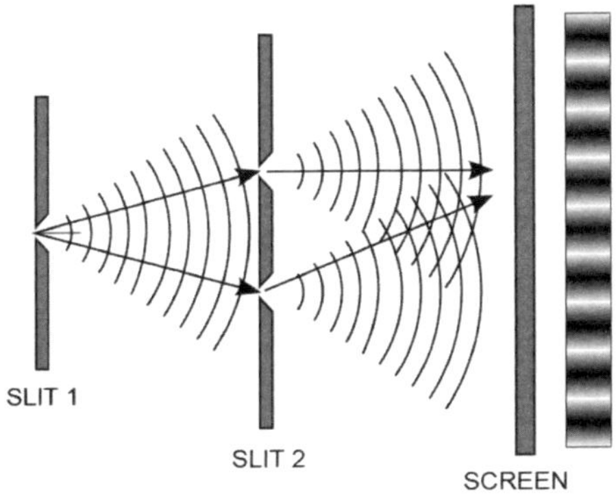

Fig 6. The 'two slit' experiment

One could question whether a creature other than human could also cause the wave function to collapse or if any form of artificial life could bring about the same result? Furthermore, is it the act of measurement that brings about the collapse or does this only occur when the mind recognises it? Although we can speculate on these points, one fact seems undisputable; *that consciousness is the instigator.*

Consciousness and the mind/body connection

If we are to comprehend more fully how and when the wave function collapse into particle reality occurs, we need to understand how consciousness, functioning through the mind-body state, operates. An often-used analogy is that of a computer connected wirelessly to the Internet. The information being accessed exists in the frequency range beyond human sight. Normally, we have no conscious awareness of this information, just as we have no awareness of radio or television signals – until we access them. We become aware of information from cyber-space when we 'log on' to the worldwide web through a computer. The programme (software) that is loaded onto the machine (hardware) reads the information by decoding it into a format that our mind can recognise, presenting it on the computer screen in the form of words, images and sounds. Our brain, which is being utilised by our consciousness through the mechanism of the mind, then further analyses and decodes this information through our bodily senses into something that we can understand. Were we to further enhance our sensory exposure by wearing a specially adapted headpiece that delivered a fuller, multi-dimensional experience it may even seem that a computer game was real to us and this would be reflected by changes in our physiology.

We can begin to see then that in a very similar way, our brain decodes information from numerous levels through the mind-body biological computer system into the illusory 'solid-world' reality that is our daily experience. The only reality is that it is no more 'solid' than the moving characters in a computer game. They are but waveform information on a software disk that is being decoded into what appears to be a real and solid state. On a much larger scale the human brain operating through the genetic structure of the mind-body biological computer transforms information from the waveform cosmic Internet or *metaphysical universe* and depicts it upon the viewing screen we each have inside our head.

A good analogy would be to consider light (consciousness) shining through a stained glass window (conditioned mind) and reflecting upon the floor (viewing screen). As it passes through the many colours within the window the white light is filtered to reflect the various hues of the stained glass and the overall image is reflected upon the floor. If we saw only the reflected image upon the ground we may consider this to be real without realising that it is but an imitation, albeit a vivid one, of that which has passed through the filtering process. Were we able to bypass the window (mind) completely and see the source of the light directly without being blinded by its intensity then our whole experience would be a very different one.

Many centuries ago, Plato (429-347 B.C.) created a useful allegory in his work *The Republic.* In the book Plato has Socrates describe a group of prisoners who have lived all of their lives chained and facing a blank wall, unable to move their heads or turn around. They observe shadows projected on the wall by objects passing in front of a fire behind them, and believe that these forms are real. Plato suggests that the shadows are as close as the prisoners ever get to viewing reality and goes on to explain how the philosopher is like a prisoner who is released from the cave, coming to realise that the shadows on the wall do not constitute reality at all, as he can perceive the true form of reality and not simply the images witnessed by the prisoners.

Returning to our previous analogy, if we consider that large numbers of computers connecting wirelessly to the Internet share information that is passed electronically through cyber-space before decoding it into a format that our brain can recognise, we may be able to see parallels with our own systems of decoding information contained within the metaphysical universe – the cosmic vibrational fields that surround and interpenetrate each and every one of us. There is no *outer* world; it only appears that way because of the way in which we decode it. What we term physical reality is *within us* and our experiences are a direct result of what we have chosen either consciously or unconsciously to decrypt from the energetic ocean of which we are a part. We should thus remain aware that the mind-body computer decoding system is *only a vehicle for consciousness* and not reality itself. Consciousness is the essence of what we are and if it were removed from the equation we would be left with just the five-sense realm of body-mind. This would be akin to programming a machine and then leaving it to decide for itself the outcome of its actions. Consciousness is the operator, the guiding intelligence that uses mind and body for expression at the physical level of creation. We are, all of us, 'in' this world, but not 'of' this world. We experience relative degrees of what we term reality and think that we understand it when this is far from true. Like the frog that spends its entire life in a garden pond, never knowing the greater world that lies beyond its small domain, so mankind catches but glimpses of the larger reality that lies beyond the frequency range of daily experience. As we peel away more of the layers and new perspectives emerge, a fundamental understanding develops – we are much more that we have ever dreamed that we are. Not only this, but *we are central to the very existence of life itself.*

Changing our perspective

Our idea of what constitutes reality is not fixed. As we are open to change, so our version of reality changes too. It can be problematic if we allow ourselves to become locked-in to particular belief systems and the way that society is currently structured ensures that this will almost inevitably occur. When a shift happens and we are able to access another level of awareness, the accompanying sense of freedom is tangible. Modern living is very much geared towards the five-sense reality and we are constantly stimulated in ways that feed our conditioned views. There is nothing wrong with enjoying the pleasures that life has to offer, but remove the blinkers, strip away the fixed ideas and beliefs and whole new dimensions open up to us.

When I write of 'decoding' information, it has to be understood that there are so many more levels of awareness with which we can and do interact. Our experience of this physical level of reality can be likened to viewing life through a hole in a wooden fence. We see a small portion of what lies beyond and recognise that there are many things there that we would like to know more about, but that we are limited whilst we remain on our own side of the fence. We only have to consider the entire electromagnetic spectrum to know that the section encompassing visual light is very small *(Fig 7)*. This being the case, it doesn't take much reasoning to conclude that we need to discover ways in which we can connect with other frequencies of vibrational reality. In short, we need to modify our decoding equipment and expand our ability to receive and interact. This need not be a difficult process and neither should it be considered an academic one requiring hours of study. A little practice is necessary, as is a willingness and desire to progress, but these elements already reside within the majority of us.

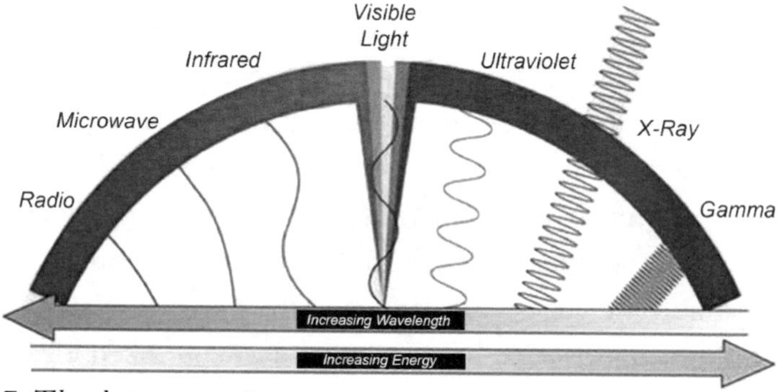

Fig 7. The electromagnetic spectrum

The holographic universe

As we begin allowing that expansion to gather momentum, let's consider another level of the decoding process and our ability to access waveform information by considering that the universe itself is holographic in nature. In 1991 a book by author Michael Talbot (1953-1992), titled *The Holographic Universe* expanded upon some of the earlier themes proposed by two of the world's most eminent thinkers, physicist David Bohm (1917-1992), a former protégé of Albert Einstein (1879-1955) and quantum physicist Karl Pribram (1919 -). The themes carried forward in the book outline the concept that the entire cosmos is a hologram, as is the human body-mind and there is compelling evidence to suggest that this is indeed true.

To comprehend what a hologram actually is we need to dive once more into the ocean of waves and understand a phenomenon known as 'interference'. As with the two-slit experiment, interference occurs when two or more waves intersect and a crisscrossing pattern emerges. Imagine for a moment, dropping a small pebble into a still pond and watching the expanding ripples move outward from its centre. Drop a second pebble into the same water and you will see the two sets of expanding ripples interfere with each other in a complex arrangement of peaks and troughs resulting from the collisions that are occurring. Indeed, any wavelike phenomena can create such a pattern, including radio and light waves. Laser light, because of its purity is especially good at creating interference patterns and with the invention of the laser it became possible to create and view holographic images. This is how it works: A single laser beam (a very concentrated form of light) is split into two separate beams by a 'beam splitter'. The first beam (working beam) is bounced off an object (an apple for example) being photographed. The second beam is allowed to go directly to the film and to collide with the reflected light from first beam. Here an interference pattern occurs and is recorded on the film. When this pattern is 'read' (decoded) by another laser beam an apparently three-dimensional image appears.

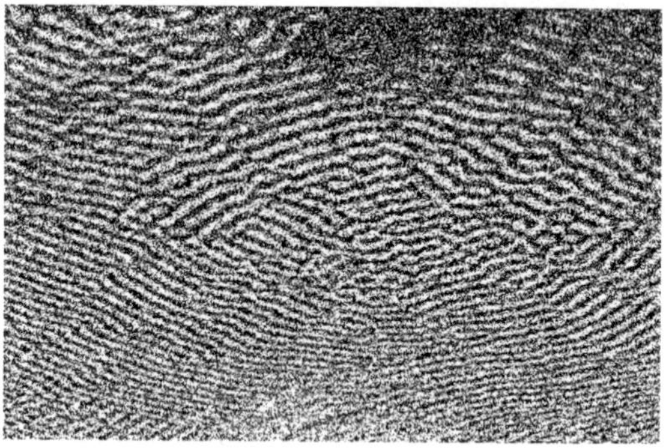

Fig 8. A typical holographic interference pattern – waveform information yet to be decoded into an illusory three-dimensional form

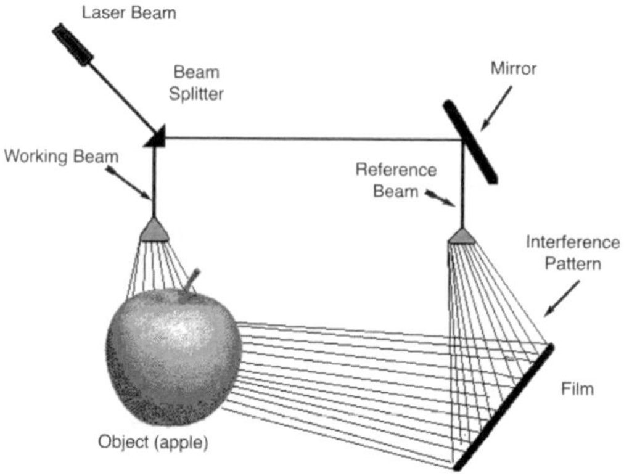

Fig 9. A holographic image produced by a laser beam

The fascinating thing about a hologram is not only that it encompasses a three-dimensional image of the object photographed but that even when the film is cut into smaller pieces, each piece still shows the whole picture. No matter how many times the film is chopped up, the entire image still appears on each portion. In essence, every part is a smaller version of the whole. Perhaps this is worth considering in relation to practices such as reflexology, acupuncture, iridology and other healing disciplines that address the whole body through a smaller part of it. If of course, the body is indeed holographic in nature then this would go a long way to explaining how such treatments work.

The correlation between holographic images and waveforms becomes increasingly significant when considered from the quantum perspective. If ever there were any doubt that the waveform and particle can exist simultaneously, (as some scientists have questioned), then the hologram goes some way to providing an explanation by way of proving that the interference pattern is the information from which the three-dimensional hologram is read. Were there no interference pattern, there would be no 3D image – one is a part of the other and their relationship is symbiotic. The information contained within the waveform is still present even when it is not being decoded, but it remains in that state in exactly the same way as our own 'reality' does before we also decode it into 'solid' form.

What we are also witnessing here, in both the hologram and the collapse of the waveform into the particle or put another way, the decoding of the information into the reality, is the introduction of *order* into a chaotic or disordered system. In the truest sense, there may not actually be disorder but rather *an order that is hidden.* Bohm recognised this and suggested that this deeper level of order was enfolded, referring to it as *implicate.* Our own level of existence he refers to as *explicate,* meaning unfolded order. Within this context our piece of holographic film and its generated image are an example of implicate and explicate order – the film being implicate because the encoded information within its interference patterns is a hidden one and the projected hologram an explicate order because it represents the unfolded or decoded version of the captured image.

Another good analogy offered by Bohm to explain a phenomenon known as *non-locality,* which occurs at the quantum level (and which also dovetails nicely with our attempts to explain the nature of how we manifest reality) is the fish-tank experiment. Imagine an oblong tank, filled with water and containing one fish. There are two cameras filming the fish – one situated at the end of the tank, seeing the fish head on and another at the side of the tank, seeing the fish side on. Two separate monitors show the fish from the two angles and as the fish moves, both images appear to move correspondingly. When one faces the front, the other faces the side and so on. If you were oblivious to the set up you might conclude that there were two separate fish that seemed to be communicating with each other, but of course this is not so. No communication is ever taking place because there is only one fish, not two. According to Bohm, this is exactly what is occurring between particles at the subatomic level and of course because of the potential within the quantum existence, all particles must be *non-locally* connected.

The concept of non-locality wonderfully describes the connection that exists between particles and goes some way in explaining the instantaneous communications that take place at the quantum level. In his book *The Divine Matrix* author Greg Braden (1954 -), quotes from an eloquent description given by Russell Targ (1934 -), a cofounder of the Stanford Research Institute in Menlo Park: *"It's not that I close my eyes and send a message to a person a thousand miles away, but rather in some sense there is no separation between my consciousness and his consciousness."* Precisely, because the signals didn't have to travel anywhere – they were already there. In the holographic model, every place is but a reflection of every other place with everything connected by a unified underlying field.

Holographic memory

Whilst considering the hologram it is also worth mentioning here the nature of memory. Again, according to mainstream belief our memory exists within the brain structure and as we grow older we lose the ability to access it to a greater or lesser extent because of disease or the ageing process. Whilst it may be true that from the physical perspective our body loses some of the attributes and vitality that it had when in its prime, the real memory is actually unaffected because it is a non-physical phenomenon, existing as a waveform throughout the entire genetic structure. I would further suggest that memory itself is holographic in nature and has many attributes unknown to science. If we consider one of the most abundant properties existing today upon planet Earth – water, as a prime example of how memory is retained, we will recognise comparisons with the human form and the way in which consciousness operates.

Our bodies contain a large percentage of water (80% at

birth) and this is very significant when you consider that water has been proven to retain information. Dr. Jacques Beneviste (1935-2004), a French doctor specialising in immunology, made several significant findings relating to the properties of water. Amongst these he discovered that when a substance is diluted in water, the water carries the memory of the substance even after it has been significantly diluted so that none of the original molecules of the substance remain. Those of you familiar with homeopathy will recognise this principle. He also found that the molecules of any given substance have a spectrum of frequencies that can be recorded digitally on a computer and then played back into untreated water resulting in the 'treated' water then acting as if the actual original substance were physically present.

Another innovator, Dr. Masaru Emoto (1943 -) also carried out research into water and came to similar conclusions. One of the experiments carried out involved freezing water and then examining the images under a special dark-field microscope. This revealed that water from a pure source such as a natural spring formed into wonderful crystalline structures, resembling snowflakes, whilst polluted water did not. Further experimentation involving the use of various words, names of people, prayers and music also yielded fascinating results. Experiments showed that the name 'Hitler' and negative phrases such as 'I want to kill you' and 'I hate you' prevented the same elegant form of crystalisation from taking place whilst positive words and phrases such as 'I love you' and 'Thank you' had a beneficial effect. In addition, the differences between water exposed to heavy metal and classical music showed marked differences in appearance clearly proving the influence of positive or negative energy. This makes perfect sense when we understand that water has consciousness within it (as does everything in degree) and has the ability to store information digitally and holographically.

The oneness of life

When the mystic or the enlightened teacher speaks of the connectedness of all things or the 'oneness' of reality, is this not what they mean? They may or may not possess any scientific understanding but they do know from a deeper level of awareness, or greater perspective, the same truth. Do not be concerned if this is a difficult concept to grasp. When we view anything from within, it can be difficult at first to gain an understanding of the whole and the old saying 'I can't see the wood for the trees' springs to mind here. Yet there is much to consider because we exist within fields of infinite possibility. If we are indeed holographic forms within the cosmic holographic reality – the metaphysical universe of all that ever was, is and will be, then we have nothing to fear for our journey can only ever be one of self-discovery. With each layer that we unfold we uncover more of ourselves and in doing so gain ever-greater insight into the source of our being. If, as I believe, we are *infinite consciousness* searching for itself within itself, upon what more glorious journey could we have embarked and what greater riches could possibly exist than to know the creator of all, *within?*

There is no separateness to anything. All is connected in a seamless whole and it is as meaningless to view the universe as composed of 'parts' as it is to view the wave as divorced from the ocean. The old conventional view of reality has changed forever and in its wake is emerging a dynamic, all encompassing vision that unites us all in one boundless continuum.

Summary of main points:

- Our true nature is individualised consciousness, itself part of infinite consciousness, often described as God, The Great Spirit, Brahma etc.
- Our brain decodes vibrational waveform information from the metaphysical universe into electrical, digital and chemical information forming a three-dimensional holographic reality
- There is a clear distinction between brain/mind/consciousness
- We are part of a shared human experience – the collective unconscious
- Consciousness is implicit in the creation of all form and reality
- At the physical level of existence we are restricted in our awareness of other frequencies but may glimpse these from time to time through various disciplines
- The entire universe is holographic in nature and everything is part of one whole

Chapter Two

The Language of Transformation

"The secret of DNA's success is that it carries information like that of a computer programme, but far more advanced. Since experience shows that intelligence is the only presently acting cause of information, we can infer that intelligence is the best explanation for the information in DNA."

Jonathon Wells (1942 -)

Having established that consciousness, operating through the mind-body system decodes information from numerous frequencies and dimensions to create 'reality' we have to acknowledge the influence that these sources have upon us, some more than others. Just as we receive stimuli through our five physical senses (visual, auditory, gustatory, olfactory and kinesthetic) so we are also influenced through other sense abilities of a non-physical nature including so called psychic capabilities of which clairvoyance, clairaudience and clairsentience are the main ones. As well as this, influences from those who exist in non-physical realms also play a part, although we may be entirely oblivious to these. We also hold a great deal of pre-programmed knowledge within our DNA structure and this, combined with the unconscious aspect of our mind and our personality plus our ego, influences our behaviour and decision-making. In addition, certain hereditary predispositions giving rise to both inherent strengths and weaknesses catagorise the way in which we develop as well as containing latent information that can be triggered later in life. It is also true that each of us possess a degree of knowingness or wisdom that is the aggregate of previous experiences or what are commonly referred to as past lives. This brings into play of course, the controversial topic of reincarnation, but more of this later. For now, let us focus upon our inherited and acquired individual and collective belief system along with current day-to-day programming, much of which occurs the unconscious level.

Because each of us have a degree of freewill to determine to a large extent what we think, say and do there exists a belief that we are more or less agents of our own destiny. To an extent this is true, because if we take the wider holistic view, we become what we ourselves create and despite the seeming inequality and injustice of life, we cannot be where we are not meant to be. In a universe governed by the unwavering principles of causality, chance and randomness cannot exist – although quantum theory may challenge this point. We are however, subject to so many influences that impact upon our freewill and colour our decision-making. Having worked for a while in advertising I am only too aware of the power of marketing and were it to be as ineffective as some would have us believe, then the vast amounts of money spent each year on various media campaigns would have no direct bearing on sales, which clearly is not the case.

We should not though, assume that all marketing is bad for us because clearly, adverts that address important issues such as health, safety and awareness of certain issues and topics beneficial to our well-being can be helpful. What is not good for us are the kind of messages, some of which are subliminal, that influence the mind in ways that are less than beneficial and in some instances, harmful. How often are people told what to eat, drink, watch, wear and think? Role models are created by agencies wishing to portray a certain style or attitude in order to sell more of a particular brand with profit frequently being the motivating factor. In addition, those wishing to push a more covert agenda may utilise the media in ways that are not easily discernable. It was not so long ago that certain television commercials were found to contain split second messages that worked on a subliminal level to impart information of which the conscious mind was unaware.

Imagery of all kinds is also regularly used to convey messages to the human psyche and Hollywood in particular plays an extremely influencial role in seeding ideas in preparation for future events. It is not my intention to delve further into the realm of conspiracy theory at this point, (although I have my own ideas on this), but it is important to point out that the human mind is subject to manipulation and that not everyone in the world has a benevolent agenda.

The power of language

Words are extremely powerful and should never be underestimated. In essence a word is just a sound, but it is the *meaning* contained within the sound that is important. If we take the word 'love' as an example of a 'good' word, we are all familiar with the connotations of kindness, affection, gentleness, warmth, happiness and so on, that are evoked when we either see, hear or speak it. Equally, when the word 'evil' is used, our mind instantly conjures up thoughts of pain, suffering, darkness, nastiness, selfishness, cruelty and all manner of negative or bad ideas and images fill our consciousness. What is interesting though is that the brain remembers the association between the feeling or emotion and the word and stores it for future use. There are good reasons for this of course, because certain words are important in that they convey an instruction or a warning about something that we need to know. The word 'STOP" for example, especially if accompanied by a visual image or the colour red that is often associated with danger, usually prevents us from going on further without at least pausing to consider what lies ahead.

Swear words are also very powerful and emotive. Many of us swear from time to time, especially if we have hit our thumb with a hammer or trodden in something particularly unpleasant when we are wearing our best shoes. A swear word can act as a release of energy, notably when we are angry and so is helpful when viewed in this way. Unfortunately, it may not be so helpful if we swear inappropriately or directly at another person and so provoke an angry response within them.

What is also interesting is that the brain will recognise certain words and their connotations *even without seeing the whole word.* For example the word B L * * * DY when written in this way is not a word at all. Yet the brain fills in the missing blanks and takes a calculated guess that is usually, if not always, correct. Certainly within the context of a sentence there would be less likelihood of an incorrect assumption. Look at the sentence

The Language of Transformation

below and see if you can read it:

M a * y   peo * le   e * joy   stud * ing   so * cial   ps * chol * gy

Here's another example:

7H15 M3554G3

53RV35 7O PR0V3

H0W 0UR M1ND5 C4N

D0 4M4Z1NG 7H1NG5!

1MPR3551V3 7H1NG5!

1N 7H3 B3G1NN1NG

17 WA5 H4RD BU7

N0W, 0N 7H15 LIN3

Y0UR M1ND 1S

R34D1NG 17

4U70M471C4LLY

W17H 0U7 3V3N

7H1NK1NG 4B0U7 17

Unless you are unfortunate enough to suffer with dyslexia or 'word blindness' you probably managed to read both examples quite easily and quickly as your brain added in the missing letters. Other simple examples of the brain recognising words include popular 'Word Search' puzzles and of course the ever popular Crossword games. More complex examples can be found within some IQ tests and problem solving assessments and often include shapes *(Fig 10)*.

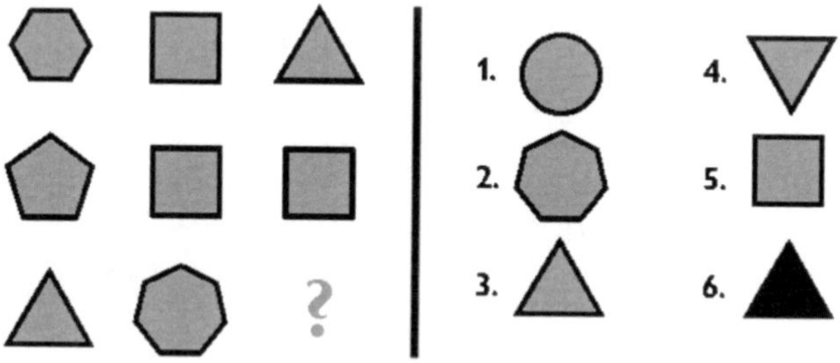

Here again the brain is being asked to work out what the next image in the sequence should be (from a choice of six options) and although the solution is not so easily forthcoming, a person with a reasonably high IQ will find the answer.

The purpose behind these basic examples is to illustrate how in a simple way the brain recognises certain forms and is able to fill in blank areas in an intelligent way when required to do so. It is then, important to remember that the use of both imagery and language when applied with purposeful intent can bring about powerful change that affects us on many levels.

Hypnotic interventions

Several years ago I studied and gained a qualification in the use of Ericksonian Hypnosis and NLP (Neuro-linguistic programming). These techniques employ the therapeutic use of certain language patterns and interventions to positively manipulate clients towards a state of wellbeing. One of the terms used to describe the Ericksonian techniques is known as 'artfully vague language', which as it suggests employs the use of certain language patterns that are deliberately ambiguous in order to offer the client a choice. Often there is no choice because of the assumption contained within the wording and phrasing of the intervention, but therapeutic terms such as *binds, double binds and unconscious double binds* nevertheless allow the facilitator to impart the illusion of choice with the assumption that one or the other alternative will be chosen. For example, to enquire "do you want to go into trance in this chair or that chair?" involves the therapist asking a simple question which can be answered consciously. However, the decision to sit in one chair or the other effectively binds the patient to the task of going into trance regardless of which chair is chosen. A competent therapist will, after having gained rapport with their client, proceed to lead them into an altered state of consciousness or *trance state* by careful and skillful use of words and sentences that they have mastered. There is a broad scope for creativity and the use of metaphor and story telling techniques are often employed to great effect later in the session when the subject is more suggestible. The main criteria for any good therapist, is their ability to bypass the subject's conscious mind and therefore address the subconscious parts directly.

Once the analytical aspect of the conscious mind can be circumnavigated and the unconscious mind accessed directly, positive suggestions can be more readily implanted and beneficial change brought about.

It is worth noting here that should a therapist suggest something sinister to a subject in a hypnotised state, which they would under normal circumstances reject – for example, that they murder someone, the likelihood is that the suggestion will not be accepted and will be overridden by their own moral compass. There are in use today though, more powerful methods of facilitating changes within the mind than those employed in a beneficial way by respected therapists and it is known that certain government agencies utilise brainwashing techniques and programmes such as 'MK Ultra' and others to obtain results favourable to their agenda. It is also known that powerful images such as the 'all seeing eye' amongst numerous others can effectively be used to programme people, often subliminally. Logos of the kind used by large corporations are particularly favoured to impart a deeper message, as are some iconic images that appear in promotional videos, especially within the pop music industry.

'Reverse Speech' is another technique that is sometimes used to deliver and implant a message without the conscious mind being aware and there are numerous examples of this to be found on the Internet. Although some people doubt the validity of this technique, I know from my own research that certain words and phrases when played in reverse, often contain quite distinct and coherent sentences. It is quite astonishing how, when we hear something played backwards it seems to bear little resemblance to the original forward sentence.

When you have knowledge of language patterns and their deployment it is possible to use your own filtering mechanism to become more aware of the implication behind the words. It is not possible to do this all of the time but with practice there is a definite sharpening of the senses and it becomes easier to pick out certain key words and phrases that are NLP based. For example, the use of *nominalisations* by certain keynote speakers, particularly politicians, can be a useful tool to try and elicit votes. The word 'change' is one that is often used. This can mean almost anything you want it to and if you asked any number of people if they desired change in their life, they would inevitably agree. The key here is that almost instantly, rapport is established between the speaker and their audience because they have offered to provide something that their public wants, even though that could be almost anything from a new pair of shoes to a pay rise. Next time you witness a political debate or a salesman knocks on your door, put on the filter and listen for the way in which a question is posed, an answer given or a promise made – you'll be delightfully surprised to discover how many embedded commands or multiple choice options that lead to only one choice are present.

Programming with words

Whilst we are on the subject, recent research by Russian scientists has revealed that DNA is programmable by language. Esoteric teachers have long since known that our body can be influenced by words, images and thoughts but this has now been scientifically proven. Human DNA acts like a biological version of the Internet and is even superior to it in many ways. DNA, or *deoxyribonucleic acid* to give it its full name, is the hereditary 'building block' of life and nearly every cell in a person's body has the same DNA, with most of it located in the cell nucleus (where it is called nuclear DNA). Smaller amounts are also found in the mitochondria (where it is known as mitochondrial DNA or mtDNA for short). Over ninety percent of DNA is classed as 'junk' - which the molecular biologists refer to as *Introns.* Introns are like huge commercial breaks or advertisements that interrupt the real program - except that in the DNA, they take up a massive amount of broadcast time! This leaves less than ten percent being used in a meaningful way to build proteins. The Russian scientists didn't buy into this though and worked with linguists and geneticists to uncover the real nature of the junk DNA and their results show that not only is DNA responsible for the construction of our body but that it also serves as a means of data storage and communication. Their findings reveal that the coding of DNA follows the same rules as human language with similarities of syntax (the way that words are used to form phrases and sentences) semantics (the study of meaning) and the core rules of grammar. They also discovered that encoded within the alkalines of our DNA are set rules, like those found within our languages. They concluded from their findings that languages did not emerge by coincidence but are a *reflection of the inherent patterns found within DNA.*

Another huge breakthrough came from an unusual

collaboration between medical doctors, physicists and linguists who discovered even more evidence of a language buried within the Introns. According to the linguists all human languages obey something called 'Zipf's Law', named after the Harvard linguistic professor George Kingsley Zipf (1902-1950). Basically this law states that within natural language utterances the frequency of any word is inversely proportional to its rank in the frequency table. Put another way it describes the probability of the occurrence of words or other items that starts high and tapers off, with a few occurring very often and many others rarely. For example; if you consider a book containing a large number of words and then count the number of times each word appears in the book you might find that the most popular word is 'the' which appears three thousand times, followed by the second most popular word 'a' which appears one thousand five hundred times and so on. At the bottom of the list is the least popular word 'Zebra' which appears just once. You then construct two columns of numbers with one being the popularity of the words running from '1' for 'the' and ''2' for 'a' until you reach '1000' for 'Zebra' and the second column showing the counts of how many times each word appeared, starting with '3000' appearances for 'the', '1500' for 'a' going down to just '1' for the word 'Zebra'. If you are keeping up so far, you then plot a graph on the right kind of graph paper, showing the order of popularity of the words against the number of times each word appears and amazingly end up with a straight line. Significantly, this straight line appears for *every human language.* The story doesn't end there though, because scientists studied DNA, (which is essentially one continuous ladder with millions of rungs and is not broken up into separate words like a book) and made artificial words by breaking up the DNA into 'words' that were each three rungs long. They then repeated this experiment for 'words' that were longer in length and having analysed their findings found the same kind of Zipf Law straight line graph for human DNA as they did for human languages.

Computer Binary Code

ATG-TCA-AAT-AAA-AGT-AAT-GAT-AAT-GGC-AGA-GCA-TAT-GAG-TTT-GCA-TTT-ATA-AAT-GAA-
TTA-GGA-CGC-ATT-GCA-ACT-CAA-AAT-CAG-AAT-ATA-AAT-ATC-GAA-AAG-AAT-TCT-AGC-TAT-
TAC-GTA-GTT-GAG-AAA-TCT-TGG-AGT-ACA-TTA-TCG-GAT-CTT-GAA-AAA-GAA-AAA-TAT-ACA-
AAA-AGT-GCA-ATT-GCT-GGT-ATC-AAT-CTT-ATA-ACA-AGC-TTA-GAG-CCA-ATA-ATA-GAA-GAT-
GGT-AAT-GGT-GTA-TTA-AAC-TTA-AAA-ATA-CAA-GCT-GAT-AAT-AAA-GGT-GAA-TTA-GGC-GAT-
ATT-AGA-GAT-ATT-TTA-ATT-CAA-AGA-GAA-AAT-ATT-AAT-TGG-GAA-ATT-GGT-TTA-AGT-TTA-
AAA-CAT-AAT-CAT-TTT-GCT-GTG-AAA-CAT-AGT-CGT-TTA-TCA-CAT-AAA-ATT-GAT-TTT-TCA-
GAA-AAA-TGG-TTC-CAA-TTA-CCT-TCT-TCT-CAA-AAT-TAT-TGG-GAT-AAT-ATA-CTC-CCT-ATT
TTT-GAG-AAA-TTA-GAA-ATT-TAT-AAA-AAA-GAT-AAA-ATA-AAA-TGG-AGA-GAG-TTA-TCT-AAT-
AAA-GAA-GAT-TGC-ATT-TAT-TAT-CCC-ATA-CTT-AAA-TCA-TTT-ATA-GCA-GAA-ATT-AAA-GAA-
AAG-TAT-GAT-AAA-TAT-AAT-TCT-ATT-GTT-CCA-CAG-AGA-ATG-GTT-GAA-TAT-TTA-CTT-GGA-

DNA Code

Fig 11. Computer binary code and DNA code

The vibrational structure of DNA is also very interesting from the point of view that it would appear that living chromosomes (thread-like structures located inside the nucleus of animal and plant cells, each containing a single molecule of DNA) "function just like a holographic computer using endogenous laser radiation" according to biophysicist and molecular biologist Pjotr Garjajev. His findings further confirmed that when certain modulated frequency patterns (sound) were incorporated onto a laser-like ray, the DNA frequency of the information was itself influenced. It was further proven that simply using words and sentences found within human language, without any further decoding, was enough to influence the DNA and it seems that if the appropriate frequencies of sound are used, the DNA will always react. Is it any wonder then, that NLP and hypnotic techniques, as with music, can cause a reaction at such a profound level?

It used to be that if scientists wished to modify a strand of DNA, they needed to cut and splice it, rather like editing a sequence from an old movie reel, yet the Russian researchers created devices that obtained the same results using modulated radio and light frequencies. They even went as far as changing frog embryos to salamander embryos simply by transmitting DNA information patterns, apparently without any side effects.

Meanwhile in the USA, bioengineering researchers at Stanford School of Medicine have found a way to digitally encode, store and erase data within the DNA of living cells. In a paper published in 2012, Jerome Bonnet PhD., a postdoctoral student and two colleagues Pakpoom Subsoontorn and assistant professor Drew Endy PhD., discovered a way to reapply natural enzymes adapted from bacteria to flip specific sequences of DNA back and forth at will. What this means in layman's terms is that they have devised the genetic equivalent of a binary digit - a "bit" in data parlance. Using computer terminology, their work forms the basis of what is referred to as *non-volatile memory* or data storage that can retain information without consuming power. In biotechnology, it is known by a more technical term, *recombinase-mediated DNA inversion* (RAD for short) named after the enzymatic processes used to cut, flip and recombine DNA within the cell. To make their system work, the team had to control the exact dynamics of two opposing proteins called *integrase* and *excisionase,* within the microbes and experimented over and over again until they got the balance right. The implications for this discovery are enormous and Bonnet has since tested RAD modules in single microbes that have doubled more than a hundred times and each time the switch has held. It has also worked in reverse when the 'latch' has been switched back, watching a cell double up to ninety times. The latch will even store information when the enzymes are not present.

Endy acknowledges the difficulty of the work but says "We're probably looking at a decade from when we started to get to a full byte. But, by focusing today on tools that improve the engineering cycle at the heart of biotechnology, we'll help make all future engineering of biology easier, and that will lead us to much more interesting places."

Connecting with our DNA

One may speculate whether it is possible for us to communicate directly with our DNA without the requirement of any sophisticated devices such as those used by the aforementioned researchers. It is my assertion that this is indeed possible and that the more highly developed one's consciousness is, the more likely that it can happen. I would further suggest that it is already occurring albeit without our conscious cooperation, because DNA is not fixed or immutable within the course of a lifetime. There exists what I refer to as a 'baseline' level within the double helix that holds fast our essential structure but additionally there are other aspects that are responsive to elements such as radiation, viruses, chemicals and other currently unknown influences. Certainly the discovery that language can have such a profound effect is something that is revolutionary, with far reaching implications.

If it is possible for us to communicate with DNA then surely it is likely that DNA itself also communicates with us and also with other levels of creation? The simple answer is yes and in their book *Vernetzte Intelligenz* (Networked Intelligence) authors Grazyna Fosar and Franz Bludorf explain how what they refer to as *hyper-communication* operates. You and I may recognise this as 'intuition' but the authors explain that DNA plays a vital role in the acquisition of information from other dimensional frequencies. If like me, you have received insights or revelations that seem to fill your mind with previously unknown information, sometimes in part and at other times whole, then you will sense a familiarity with what they are describing. It seems that with or without our conscious intervention, large amounts of information previously unknown to us can be transmitted, even in the sleep state, resulting in measurable changes within our DNA. If you have ever had a dream in which you were communicating with a source, human or otherwise, that was imparting information to you and when you awoke you had a clear or partial recollection of something profound occurring it may be that you were accessing data in a non-linear, non-physical way. Spiritualists may ascribe this to 'visiting the astral planes' or 'meeting their spirit guide' whilst 'out of the body' and this may well be the case, but the research underpinning hyper-communication suggest that it may not be the only explanation.

Other phenomena often blamed upon the energies emanating from psychics, mediums, and healers, such as electronic devices that malfunction; lights that blink on or off or TVs that change channel may also be part of this same effect, with strong electromagnetic fields frequently generated by such individuals. Such occurrences do not involve any supernatural intervention or help from the spirit world. Again, Fosar and Bludorf suggest that DNA is the culprit and many controlled experiments have revealed an effect known as the *Phantom DNA Effect* or PDE as I will refer to it. Our old friends the Russian scientists irradiated DNA samples with laser light (remember this from our discussion on holography?) and on a screen a typical wave pattern was formed. When the DNA sample was removed the wave pattern did not disappear but remained in place. Subsequent experiments (the bedrock of the scientific method) revealed that the pattern continued to come from the removed sample, where an energy field apparently remained intact. The suggestion is that energy from beyond time and space still flows through 'activated wormholes' after the removal of the DNA, although this requires further research in my view, but anyone reading this who has ever experienced 'phantom limb pain' will perhaps appreciate this explanation.

If your memory is holding out with all of this fresh information that you are absorbing (and don't forget I told you right at the start that we'd be moving along at quite a pace) you may recall that we touched briefly upon what I referred to as the 'collective unconscious'. It probably won't come as any great surprise then that it may be quite feasible to create our own group consciousness within that domain – a kind of social networking sub–progamme if you like, where we attain access to great swathes of information via our DNA. This would be similar to 'logging on' to the worldwide web and visiting a particular site to obtain information from the vast network of material available. It is now known that just as we are able to usefully employ the Internet so our DNA can both input and retrieve data as well as establishing contact with others with whom we network. Perhaps some disciplines such as telepathy, remote viewing and even various forms of distant healing that involve the transfer of energy can be explained through the concept of group or collective consciousness and hyper-communication? Imagine what could be achieved if individually and collectively, humanity were to regain full control over abilities that have become somehow diminished over time?

Another researcher, Dr. Glen Rein found that negative emotional states impacted upon DNA, causing it to contract, whilst positive emotional states caused it to unwind. This would also seem to fit in with Dr Emoto's experiments with water, mentioned in the previous chapter and it is of course well documented that positive mental and emotional states have beneficial effects upon the body, making perfect sense that at the deepest levels within us, consciousness should impact in this way. Imagine for a moment a 'group consciousness', whose members, having developed and synchronised their thoughts to a higher level, focused their intent in a laser-like way upon some issue that needed resolving, such as a conflict or environmental problem.

The intent alone, behind such an initiative, would be extremely powerful. Did not some of the indigenous tribes of the ancient world wield a power over the natural elements of the earth? True, the red man may have not understood the nature of DNA but he just as surely drew upon its power when engaging in a rain dance or summoning the Great Spirit to provide for his needs.

The universal code

We've spoken of 'decoding' the information contained in waveforms and described how the brain converts this into the seemingly real world that we experience. But what about the information that is held within the genetic codes of DNA, how do we access this? If we consider that we are biological computers, then DNA is an integral part of the hard-drive that runs the machine. In *(Fig 11.)* there is an example of a typical DNA sequence showing the four codes known as A, C, G, and T (Adenine, Cytosine, Guanine and Thymine). The actual arrangement of these codes determines whether something manifests as a human, a rabbit, a leaf or a virus for example. The actual differences in code do not have to be enormous in comparison to the form itself, with just small changes creating huge differences. A chimpanzee for instance, has around ninety-six percent of human DNA and a Gorilla even more at over ninety-seven percent. The DNA of a human consists of more than three billion base pairs of these four acids and in computer terminology this equals about 750 Megabytes of data. This holds all of the instructions needed for the development of a complete living organism and is often referred to as the 'blueprint' to construct all cells, organs, skin, hair, nails and so on. This genetic information is stored in segments known as 'genes' and at the physical level this is the main purpose of DNA.

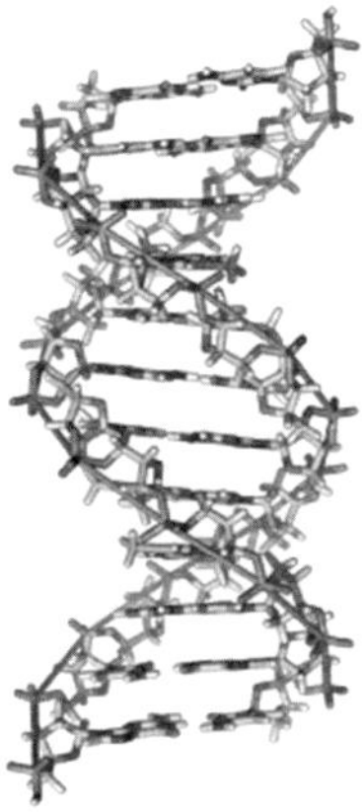

Fig 12. The characteristic double helix of the DNA molecule

If we also liken DNA to a software programme that offers infinite consciousness the opportunity to experience itself as 'human' we can appreciate that this information provides the equivalent of computer binary-code, which when decoded or 'read' enables the machine to function in a particular way.

The question naturally arises here 'read by what?' The answer is *ribonucleic acid* or RNA as it is commonly known. This single strand nucleic acid has many roles, but acts as the means to decode the information stored within DNA, just as a laser reads the information on a DVD or Blue Ray disk or as a laser light reads (decodes) the information held within the waveform of the hologram.

Some researchers, notably Dr Robin Kelly (1951-) in his book *The Human Antenna: Reading the Language of the Universe in the Songs of Our Cells* - have further speculated that DNA does indeed behave as a 'human antenna', both transmitting and receiving information whilst being crystalline in structure. The Earth is also enormously crystalline which comes as no surprise as it too is continuously acting in the same way. Kelly claims that we are genetically coded with the power of *resonance* with the ability to connect to both physical and metaphysical realms. He suggests that the blueprint for resonance resides in the coiled double helix shape of our DNA *(see Fig 12.)* whose shape acts to create a balance between opposing magnetic forces, much like the caduceus, which allows 'another dimension of energy to enter its coiled structure.'

We could consider the DNA antenna as being like a *mobius coil.* When an electrical current is passed through a wire that is coiled like a mobius loop *(Fig 13.)* the positive and negative currents balance each other out, producing a vacuum outside of the electromagnetic field in a similar way to the 'eye' within the centre of a hurricane. This space has been likened to the opening to another dimension – an energy existing outside our concept of space-time and often referred to as zero-point and scalar energy. Because the connective tissues within the human body are filled with these double helical structures, this implies we are capable of accessing higher dimensions when we are in a state of balance or attunement. Furthermore, this is suggestive of the fact that *spirituality is encoded within human genes.*

Fig 13. The Mobius Loop

Contemporary cell research, based upon the principles of quantum physics shows that the world is being created out of energy and research has demonstrated that cellular membranes contain special proteins called *Integral Membrane Proteins* that respond to energy signals from the environment, including thought. When we are able to close down our perpetual mind-chatter and go within, we align ourselves with this subtle quantum-level energy matrix and when this energy is allowed to transfer to our DNA without being interrupted by negative elements it can affect the molecular and cellular levels that drive all of our physical metabolic processes.

The body is also constantly generating its own energy waves and it has been speculated that so called *Scalar waves,* otherwise known as *longitudinal waves* can be created within the production centres of our cells. These waves, which are produced when two electromagnetic waves of the same frequency are exactly 'out of phase' (opposite each other) resulting in the amplitudes subtracting and cancelling each other out, purportedly travel faster than light. Scalar waves are often referred to as 'Tesla waves' after the scientist Nikola Tesla (1846-1943), whose research led to many wonderful inventions that paved the way for modern civilisation. Although I acknowledge that some of the research surrounding scalar waves remains unproven, it is worth further consideration here due to its possible relevance in the way that DNA acts as a receiving and transmitting medium.

The DNA antenna within our mitochondria assumes a shape known as a super-coil and these have the appearance of a series of mobius coils. It is this configuration that is hypothetically able to generate scalar waves and most of the cells found throughout the body contain many thousands of these *supercoils.* They are found within our vascular system amongst others and it no coincidence that the circulation of the blood throughout the body resembles the figure eight shape of the mobius coil. But the key here is that Scalar Waves generated in the body provide an *energetic communication system of enormous potential, connecting all cells.*

Scalar waves have many unique properties that have been tested extensively, including being able to penetrate any solid object thus sending power to a receiver with which it shares a resonance. If indeed these waves are also capable of *superluminal travel* (faster than light) as has been claimed, then the potential for all kinds of applications would be unlimited.

Mind chatter and repetitive thinking

By now you may be starting to recognise the links between language in its various forms and the influence it has upon our genetic make-up. As we often 'think' in words, it is logical to assume that our own thought processes and internal self-talk also have an impact on a deeper level and this is indeed the case. Our mind is always active, processing information from numerous sources and varying levels through the computer brain.

Whilst all of this is taking place we may often be aware of a kind of running commentary taking place, either upon the periphery of our awareness or central to it, depending upon our circumstances. Rather like breathing or some other essential bodily function occurring beneath the threshold of conscious awareness, our self- commentary is ever present but we are not always aware of it. When awareness is present we may notice that what seems to be 'an inner voice' is engaging with us, often with accompanying emotional feelings.

The mind is prone to repetitiveness and habitual modes of thought, many of which centre around anxiety, worry or concern. What I refer to as *addictive-thinking* is a process occurring within every human psyche in varying degrees and happens frequently throughout the waking hours. A typically repetitive thought cycle could encompass anything from humming a tune over and over again, a trivial concern or worry about something that the mind has identified as a problem – even though it may not actually be problematic in itself, to something more serious that is a definite threat. This ever present commentary is what I call *mind chatter.*

Examples of trivial repetitive thoughts:

"I just can't get that tune out of my head"
"What was the name of the actor in that film?"
"What can I have for dinner this evening?
"I'd better check to see if that email has arrived yet."

Examples of addictive thinking perceived as problematic:

"What if my football team is relegated this year?"
"What if she wears the same dress to the party tonight?"
"I don't know why he hasn't phoned me today."

Examples that addictive thinking related to a genuine threat:

"I am so worried about those test results"
"What will happen if she says no?"
"If that money doesn't come through by tomorrow I'm finished."

These random examples illustrate that much of our mind-chatter is centered around events, *real or imagined* that may or may not occur in what is thought of as *the future.* In addition, what the thinker is considering could happen in the future is also often correlated to thoughts of what did happen in what is perceived as *the past.* Therapists refer to this as *'past remembered, future anticipated'.* Thus a *timeline* is created within the mind, linking past and future, often without any consideration of the current moment or 'now'. In addition, the object of the commentary is frequently, although not exclusively related to the 'I' or what the mind thinks of as *the self.* This commentator is undoubtedly the egoic mind because the true self, *infinite consciousness*, is not concerned with dualistic concepts – to it, all things are one. It has no need to comment at all.

Both mind-chatter and addictive thinking are interrelated and aspects of each arise within the other. If for example there is the tendency to dwell upon a particular ailment such as a painful shoulder, the addictive element of mind will ensure that a continuous stream of thoughts related to the condition arise and mind-chatter will add a running commentary. The commentary itself, which nearly always takes the form of an internal voice, is often enough to ensure that repetition kicks in and the vicious cycle commences. The mind also incorporates other components that further compound the cycle. These are *analysis* and *emotion.* Analysis is the mind's attempt to make sense of what is going on and to understand it, for above all the mind has an insatiable desire to know things and an ignorant mind is seldom a peaceful one. Emotion, some would say, arises not from the mind, but through the mind. If we are possessing of an emotional body, sometimes referred to as the *astral body,* then it could be stated that emotions arise from this level. However, the mind has to deal with all emotions because like water carried between the banks of a river, emotions course through the channels of the mind, sometimes flooding it with their power and momentum. If we add all of these elements into the mix, we get a cycle that runs something like this:

'Mind-chatter – thought – analysis – emotion - mind-chatter'

Here then, we have another process in operation that has direct implications upon many levels of being and particularly, as we have discovered, our DNA. Yet, we should not think that all mind-chatter is negative because it isn't and where a more profound connection to our higher self is established we are more likely to receive a positive input from these non-physical levels of consciousness. Bearing in mind what we have also discovered about the human collective energy field or *collective unconscious* that I referred to in the opening chapter it is clear that the energy of language has a significant impact upon the way information is passed between us.

Language as a manifestation of consciousness

Shared language is clearly consciousness at work, and where there exists a commonality between one or more of these information fields or they become coherent and resonate in phase a new information field is created. No doubt many such fields interact with no discernable end product, but when many such fields combine, especially where there is a recognisable intent or a highly emotionally charged energy such as fear or love then this can be measured. The Global Consciousness Project (GCP) uses electromagnetically shielded computers located throughout the world that generate random numbers. In its thirteen-year history these computers have shown statistically significant deviations during global events affecting humans. For example, GCP discovered that for almost sixteen hours before the first plane struck the World Trade Centre buildings on 11th September 2001 there were 'spikes' recorded by their systems, suggestive of foreknowledge of the attacks before they actually occurred. Although counter claims have suggested that the data from the GCP computer network is inconclusive and does not show evidence of a global consciousness, some researchers believe the findings are important and would seem to fit in with the theory of *non-locality* and *quantum entanglement.*

Interestingly another high profile researcher, Rupert Sheldrake

(1942 -), coined the term *'Morphic Field'* to describe the dynamic behind this kind of connectivity and hypothesised the existence of a field within and around what he termed a 'morphic unit' (an individual form). His explanations of the nature and operation of this field bear some striking similarities to the theory of scalar waves. According to Sheldrake the 'morphic field' underlies the formation and behaviour of morphic units and can be set up by the repetition of similar acts and thoughts (intent). His hypothesis states that a particular form belonging to a certain group that has already established its collective or group 'morphic field' will tune into that field and read the information contained within it through 'morphic resonance'. To do this it must be *in phase* with it. Whether or not we choose to use the term 'morphic', it seems clear that a resonant field (or a series of such fields) exists. Such fields are by the way, not just confined to humans but also operate at lower levels of the evolutionary scale. At every level it appears that the morphic field gives each whole its characteristic properties, interconnecting and orchestrating with the constituent parts in accordance with the requirements of that level. Anyone who has ever 'owned' a pet such as a cat or dog will in all probability have experienced one or more incidents of the animal sensing something about to occur. This could be anything from 'knowing in advance' of the owner's impending trip or return home to something more serious such as a passing. Indeed, Sheldrake himself has cited this kind of behaviour as a possible effect of the morphic field. It may also shed new light on the phenomena of telepathy.

Morphic fields could be used to describe the synchronistic

behaviour of flocks of birds or shoals of fish both of whom seem to respond swiftly in unison, responding as if a single consciousness. At one level, they are. *(Fig 14).*

Flock of birds

Shoal of fish

One of the most controversial aspects of morphic fields centres on memory. Their structure appears to be determined by what has gone before, suggesting that through repetition, the patterns they organise increasingly result from probability and are habitual. The term 'habit forming' may take on an ever greater meaning here for whatever the origin of the morphic field, it seems to grow stronger through repetition with each field containing a cumulative memory that strengthens over time.

Language of the soul

Perhaps it is about time that we began to expand our rationale still further by considering the nature of our higher states of being and what I refer to as *the language of the soul.* Without fear of contradiction I will state that infinite consciousness in its pure state is not confined to language. Only in its operation at the 'physical' levels of being does it express through other forms in order to communicate. Consciousness, our core self, is vibrating way faster than anything we can conceive, resonating at truly enormous speeds in comparison to what we have any perception of. Yet because at this level of 'physical' existence or frequency range, there has to be a connective mechanism or interface through which it can register, so it utilises the mind-body computer as a means of expression, although here its luminosity is somewhat diminished and the intensity of its radiant light rather less pronounced. Like the light shone from a torch into the night sky, the farther from its source the beam travels, the dimmer it appears - *although its intrinsic properties remain constant.*

So at this our current level of awareness, the entry of the greater into the lesser, the expression of the infinite within the finite, the unraveling of embedded code into the illusion of form and the collapse of the wave into the particle are employed to great effect. Above and beyond lie the various vehicles of expression, each existing within its own plane and field of energy and each capable of influencing us from on high. *We are ourselves, experiencing ourselves within ourselves.* Within the complexity of creation run the threads of the infinite, weaving ever greater and more intricate patterns that await discovery and up until this point we have detailed but a few of the key components that play fundamental roles in determining our perception of reality and connecting us all together as one. As you would have expected, this story can never be quite so simple and straightforward, so let's journey a little further as we enter the realms that lie in the frequency ranges beyond this physical universe.

Summary of main points:

- Language and imagery play a key role in determining our perception of reality
- The brain often 'fills in the blanks' from memory
- Language is very powerful and can be used to great hypnotic effect
- DNA is a coded language that is responsive to words
- Human language emerged from the language encoded within DNA
- DNA is not fixed, but is programmable and responsive to many sources
- DNA acts as an antenna, both receiving and transmitting information
- The mind 'thinks in words' and is habitual in its thought patterns
- Resonant fields exist both within and around all living forms allowing individuals who share a common signature to communicate and share information. Distance is no barrier to this communication
- We receive information from our higher states of being

Chapter Three

Quantum Leaps

"Our minds are thus tuned or tunable to multiple dimensions, multiple realities. The freely associating mind is able to pass across time barriers, sensing the future and reappraising the past. Our minds are time machines, able to sense the flow of possibility waves from both the past and future. In my view there cannot be anything like existence without this higher form of quantum reality."

Fred Alan Wolf (1934 -)

OK, things are about to get a whole lot more fascinating as we journey into realms, or as I prefer to call them 'frequencies' that lie beyond the range of our physical senses. For some, these dimensions may be referred to as 'the afterlife', whilst for others 'spirit world' might feel more comfortable. The follower of more orthodox ways might only be content with terms like 'heaven' or 'nirvana' whilst the skeptic may feel too uncomfortable to even consider their existence.

In terms of weirdness, we have already paid a brief visit to the quantum worlds, where protons, electrons, neutrons, quarks, fermions, leptons, hadrons and yes, even bosons of the Higgs variety exist but we will be returning there as our journey unfolds and our understanding expands.

If we follow the principle of 'as above, so below' we can assume that the same basic processes apply 'up there' as 'down here' and that we manifest reality on other higher frequencies in a similar way that we do at this level. It seems logical to think that because this world appears solid enough to us who are here as creators of it, any higher frequencies, although intangible to us now, would appear equally as solid and real when our consciousness expresses at that particular frequency and sure enough, that is the case. Solidity in any event, is illusory and is but the result of resistance between interacting forces.

I am suggesting then that the same process of decoding waveform information into 'reality' continues, certainly within the immediate realms or frequencies in which we find ourselves after our consciousness has ceased expressing at this level of existence. – commonly referred to as 'death'.

This passage from the 'physical' universe to what lies beyond is but a natural transition involving the withdrawal of consciousness and despite being feared by countless numbers of people, is an essential part of life. Although there is a transfer of energy and information at the point of death and for a short while afterwards, it should be not be thought that any new vehicle of expression is being prepared or created for entry into another world at some future point. Rather, the higher vehicles of being *already exist* upon and within ever more refined frequency ranges, having been created by our own consciousness and modified along the way in accordance with the development of our character and spiritual qualities.

It is not the purpose of this book to enter into any religious debate concerning the nature of belief about the after-life, as there exist almost as many opinions as there are people to express them. I will though, offer my theories based upon many years experience as a medium in contact with the non-physical frequencies and the knowledge gained during this time, which I believe is accurate.

Considering that the electromagnetic spectrum *(Fig 7.)* is relatively small in comparison to the infinite potential number of frequencies in existence and that the range of visible light to which we are exposed is but a small percentage of this whole, it is clear that the gulf between what we actually observe and what remains for the most part undetected by our normal range of senses is enormous. Yet just because we cannot demonstrate awareness of something under normal circumstances, this does not mean it does not or cannot exist - science, as well as mediumship has demonstrated this down the ages.

For the purposes of familiarity I will also refer to these frequencies as 'planes' in some instances, whilst also retaining the use of the names to which they are commonly referred in some spiritual texts. These are however, only descriptive labels and in real terms these levels should not be thought of in isolation anymore than notes upon a keyboard. Rather they should be considered as gradations of the whole. Here, represented in a simple form are the essential non-physical frequency ranges lying immediately beyond what we refer to as the 'physical' domain *(Fig 15)*.

When considering these frequency levels, we should keep in mind that each constitutes a *level of awareness and perception* that we, as infinite consciousness, experience. We exist within and upon each of these levels *now* and are able to access and decode information from every frequency, depending upon our state of awareness. Some for example, will only be consciously aware of decoding the 'physical' level whilst experiencing life on earth, yet the more advanced amongst us will have access to information from higher frequency ranges, often in the form of insightful dreams, premonitions, inspirational guidance, past-life memories or some form of message containing knowledge and wisdom.

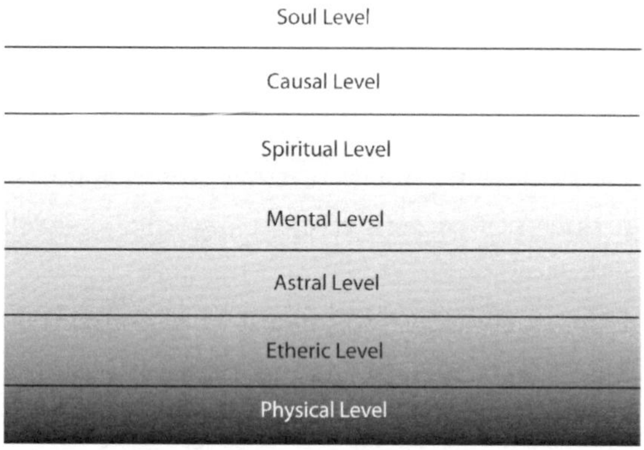

Fig 15. The planes of existence

At our current level of awareness, these other frequencies appear somewhat nebulous and unreal, falling within the domain of spiritual belief, yet just as we may be oblivious to other radio stations broadcasting on different wavelengths whilst listening to our favourite channel, they still exist, awaiting our attunement to them. When we are tuned in, it is the previous station that we can no longer hear and that seems a distant memory to us.

From what we know through more direct connections with the lower of these other dimensions, their inhabitants still retain bodies with faculties and properties that mirror to a large extent our own physical attributes. Personality persists, as does gender and the sense of individuality. Bodies appear real and solid, with limbs, sensory faculties and internal organs. There are many communications that verify these facts, most of which fall outside scientific rationale, but which an open mind will not dismiss so lightly:

"The spirit body is, broadly speaking, the counterpart of our earthly bodies. When we come into the spirit world we are recognisably ourselves."

(Monsignor Robert Hugh Benson, *Life in the World* Unseen, page 153.)

"How does the spirit appear anatomically, you will ask? Anatomically, just exactly the same as does yours. We have muscles, we have bones, we have sinews, but they are not of the earth; they are purely of spirit."

(Monsignor Robert Hugh Benson, *Life in the World Unseen*, page 153.)

"My present body is made of chemicals, and gases, and atoms - atoms certainly of a finer kind than one gets on the earth-plane. These are held together in much the same way as the atoms of the physical body, but this body does not disintegrate in the same way as the physical one does, because the life on the third sphere is sure to be as long and perhaps many times longer than the one on the earth-plane."

(Claude H. Kelway-Bamber, *Claude's Book*, Page 62)

"We have a much better-wearing body than the physical one we leave behind. It is composed of different elements from the physical one and does not need either food or sleep to replenish it. In appearance it is very much the same, but fresher looking. Lines and blemishes disappear, or rather are not apparent on the new body. Actually it is not new; we have had it all the time."

(Gordon Burdick to Grace Rosher, *The Traveller's Return* Page, 41)

From these communications, received through the resonant field known as mediumship, it can be seen that the residents of the higher frequencies appear to have retained many of their former faculties and as I say, the same mechanism with which we decode and create reality here in the physical universe still applies. Waveform information, although at much higher frequencies, is still being accessed and collapsed into 'reality' by consciousness. This is why people, places, scenery, animals and objects still appear solid and real at the level from which this occurs and at which they exist. You will notice that the quantum field and atomic particles are still in evidence and this makes perfect sense when we consider that this is still energy in operation but at considerably greater velocities. Just as we here on Earth manifest our reality, individually and collectively, so the inhabitants of the non-physical planes closest to us do the same.

I am not saying that this process is retained at much higher and finer vibratory levels where pure consciousness needs no energetic body through which to manifest, but certainly at the correspondingly lower frequencies closest to the physical spectrum, this still obtains. For ease of understanding the illustration below *(Fig 16.)* shows each body as a recognised human shape and it will be seen that as consciousness expresses through the higher and finer forms, they appear less dense as atomically their structure becomes much more refined and pure. There are variants on this model according to different belief systems, but for the purposes of this book I have retained the one with which I feel the most comfortable. I will say though, that even within each level there lie further sub-levels, much like here on Earth where there exist enormous variances across all aspects of being. If we take the astral plane as an example, there are many sub-planes existing within this frequency band and although I will refrain from speculating on their number, we can reasonably state that these are numerous and vary from the denser 'lower planes' to the finer and more subtle 'higher planes', each corresponding to the mental and spiritual states of their inhabitants. The grosser levels will naturally reflect the lesser evolved consciousness, whilst the higher levels mirror the more evolved consciousness as here again, the principle of the self-creation of both individual and collective reality discussed earlier, applies.

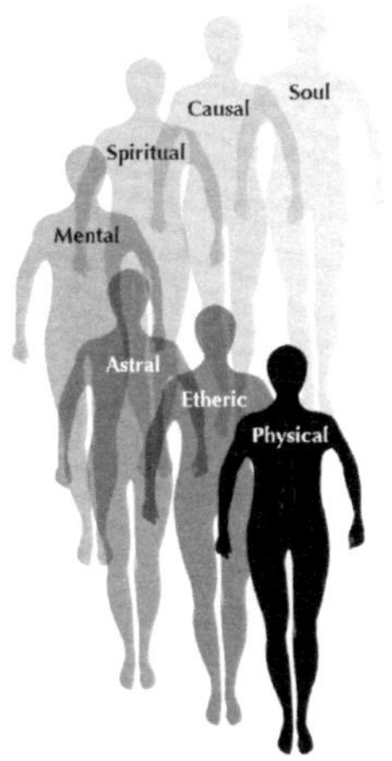

Fig 16. The vehicles of existence

As we progress both individually and collectively we become increasingly aware of other dimensions and our horizons expand to include more subtle, yet vital fields existing both within and around us. This is shown in as an ever-widening funnel *(Fig 17.)* and has enormous implications for consciousness, as we shall see.

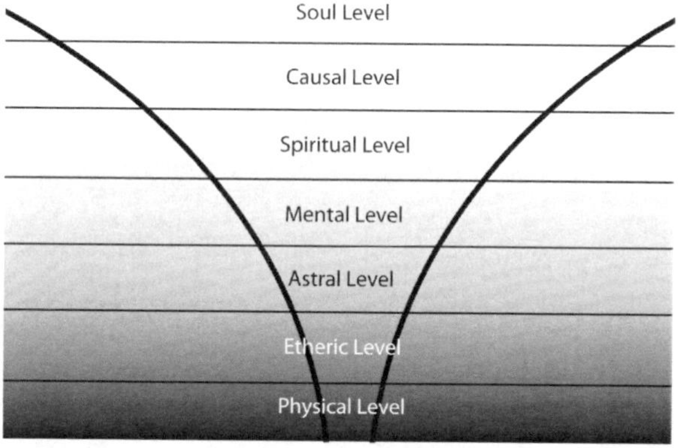

Fig 17. The funnel of expanding awareness

In terms of understanding these levels and their functionality, we need to consider both the planes themselves and also the way that we exist/decode and interact within them. What is the significance of so many frequencies? What purpose does each serve? What vehicles of expression are required to function within each dimension? These are deep and fundamentally important questions that are perhaps best understood when viewed from two main perspectives: the *personal* and the *impersonal.* The personal perspective as it relates to us - the individualised aspect of infinite consciousness (form), and the impersonal aspect as it relates to the entire nature of the plane itself (field). If we look at each plane individually, starting with the physical as the lowest and working upwards in frequency we can piece together a summary of how each corresponding level has a distinct purpose and role to play in allowing consciousness to operate.

Physical Level - The lowest and most condensed level, known as 'matter'. The physical level is a place of contrasts where opposites offer a multitude of experiences, some of which are considered 'good' and others 'bad' according to individual levels of spiritual awareness.

The physical body provides a solid foundation for all the other bodies. Experience is decoded through the mind/brain & five-sense connection. The physical body is unique to each lifetime and reverts to its atomic level at death. The five physical organs of action are hands, feet, digestive system, throat, and genitals. The physical level is composed of the energy states of solids, liquids, and gases and dependent upon the etheric body for its vitality, life, organisation, and many processes that result in wellbeing.

Etheric Level - This is the frequency band located just above the physical. Its most important function is the transfer of subtle energy from the *universal field* to the *individual field*, and thence to the physical body. It acts as a connecting link between the physical body and the astral (emotional body) and mental body and furnishes the basic pattern according to which the physical body is constructed. It is often viewed as a bluish-grey energy extending between 20mm and 100mm beyond the boundary of the physical form.

Astral Level – As the frequency band located just beyond the etheric, the astral level impacts upon the astral body through emotion - thus astral energy is often emotionally linked. It is to be noted that there are several astral planes of varying degrees and it is to a plane closest to our own physical level that we travel in our dreams and project ourselves during 'astral travel'. When seen or photographed the astral field appears to extend from 300mm to 500mm beyond the physical body appearing as a somewhat egg-shaped luminous cloud of ever changing colours.

Mental Level - This is the level located just beyond the astral frequency, but interpenetrating both the lower astral and etheric bodies (levels). It is this mental dimension that is in constant interplay with every aspect of our personality and its energy permeates every experience, even when we are not partaking in intellectual pursuits or engaging in conscious thought. The mental body extends about one metre beyond the periphery of the physical body and like the astral, it too is ovoid in shape but much larger and more refined. Its colours reflect the individual's interests and mental powers, whether latent or active.

Spiritual Level – Part of the higher soul structure, of a finer vibration than the astral and mental fields and of which we are seldom fully conscious, the spiritual level nevertheless makes up part of our personality and it is from here that we access aspects of our higher wisdom. This spiritual field can also be said to be more permanent in expression in that it is closer to the absolute than any of the aforementioned fields and therefore less open to fluctuations of emotional and mental energy. When considering more traditional terminology this level could be thought of as the true 'spirit world'.

Causal Level– Often referred to as 'the temple of the soul' and termed 'causal' because, according to esotericism, it carries our fundamental intention to *be* - the ultimate cause of our existence. The causal body is creative and intuitive, the holder of all wisdom and the preserver of karma. At this level, there are no constraints of time and space and causality, and we are able to experience the universality of life, perceiving deeper meanings that are normally not available to us. Information comes to us more easily at this level and in a non-linear way, imprinted upon the mind as a complete whole. This body does not disintegrate after the death of the physical body as the astral and mental bodies eventually do but persists from one incarnation to the next.

Soul Level – Soul is a term rarely used with precise definition in philosophy, but is the essence of who we are as an individual aspect of infinite consciousness. The soul level is the purest and most refined of all levels, being the closest to *the absolute* and can be defined as an individualised aspect of the Godhead.

The astute reader will have noticed at this juncture a pattern emerging which points towards a movement of consciousness in the direction of ever higher and finer interactions - from the gross to the refined. Each level or frequency has a vibrational link to the others and they interact with each other and with the consciousness that is our unique expression. Just as we are not separate from what we manifest as physical reality, even though it appears to be 'outside' of us, so neither are we separate from any of these vibratory states or what appears as reality within them. They exist because we exist and both are manifestations of the same consciousness. We do not for instance 'enter' the astral world upon 'leaving' the physical reality – we are already there, as we are too within every other level to which our consciousness vibrates. The 'bodies' illustrated in *(Fig 16.)*, are pulsating energy fields, interactive wave patterns of higher and finer degrees, operating within certain defined frequency ranges. What occurs when we 'die' is that our consciousness stops decoding its 'reality' within the physical range of vibrations but continues doing so within the next frequency band. There is no 'death' in this sense, just a shut down of the body-computer hardware (physical body and brain) allowing our consciousness to continue functioning through a more refined body-computer (astral body and brain or higher). I will emphasise that there is no waiting period, judgement day or 'purgatory' for anyone following the passage from one level to another. The transition for most is a seamless and continuous one and should not be feared.

It is worthwhile stating here that other aspects of our consciousness, already operating on the higher planes can also be accessed under certain conditions. We do for instance, often visit the 'astral reality' during sleep and bring back our experiences to the waking state in the form of dreams, some of which appear bizarre due to the fact that the brain cannot always process smoothly the information picked up from a higher frequency.

Both mental and physical forms of mediumship are examples of how our higher consciousness is contacted and communicates information to us and those familiar with the concept of OBEs (out of the body experiences) and NDEs (near death experiences) will appreciate that here too, consciousness is utilising a higher body as a means to experience beyond the physical. Science has documented many 'out of body' cases that prove that this happens. Over a two-year span during the 1990s, researchers Dr Kenneth Ring and Sharon Cooper completed a study into the NDEs of the blind and published their findings in a book entitled *Mindsight* in which they documented solid evidence of over thirty cases in which blind people reported visually accurate information obtained during an NDE. One such case involved Vicki Umipeg, a forty-five year old blind woman who was born blind due to her optic nerve having been completely destroyed at birth because of an excess of oxygen in her incubator. Following an automobile accident many years later, Vicki found herself floating above the (physical) form in the emergency room of a hospital and was aware of being up near the ceiling watching a male doctor and a female nurse working on her body. She told Dr. Ring of her clear recollection and how she realised that she was looking at her own body from an elevated position. It is clear from the transcript that Vicki must have moved further beyond the confines of the hospital building as she continued to relate a journey that involved being sucked headlong into a tube accompanied by feelings of being pulled upwards towards a light (this is a common feature within NDEs). She could hear music and found herself lying on grass surrounded by trees and flowers. There were many people around her and she described a tremendous light that could be felt as well as seen as she began to become aware of friends she had known, but not visualised, whilst on earth. The transcript continues to include descriptions of two of her childhood caretakers, two other blind schoolmates who had died many years earlier at ages eleven and six and a married couple with the surname of Zik, both of whom had previously passed away. The culmination of the transcript seems to centre around a telepathic communication from a celestial being who imparted a gentle message telling her amongst other things, that she had to return to her physical body and continue her earthly life – which she did with 'a sickening thud'.

This type of experience is not uncommon and most of us will have had watered down versions of our own when we have experienced lucid dreams of other dimensions and interactions with people we have known that may have passed on. How often have we also felt the sensation of returning to our physical body with a start and suddenly waking up with a deep sense of knowing that we have been somewhere else during our 'sleep state'?

Recently, the most astonishing evidence has also emerged in a book titled *Proof of Heaven* by Dr. Eben Alexander III (1953 -). Working as an academic neurosurgeon for over 25 years, including 15 years at the Brigham & Women's Hospital, the Children's Hospital and Harvard Medical School in Boston, Alexander specialised in helping patients rendered comatose by trauma, brain tumors, ruptured aneurysms, infections, or strokes and thought he had a pretty good grasp of how the brain and consciousness operated, believing like many mainstream practitioners, that the latter was a product of the former. That was until he too was unexpectedly driven into coma by a rare form of bacterial meningitis-encephalitis. With his brain almost totally inactive and its neo-cortex (the part of the brain responsible for all our thoughts, emotions and everything that makes us distinctly human) completely dysfunctional, scans revealed the massive damage that had occurred. He was not expected to survive, but survive he did and after seven days miraculously emerged from the comatose state as a changed man. Prompted by his son to recount his remarkable experiences Alexander pieced together the notes he was able to make as soon as he had regained the ability to write and detailed what had taken place during his NDE. The story of his survival is compelling, especially as his experiences whilst in the coma cannot be simply explained away as a dream, hallucination or any other mental phenomenon, but what is so strikingly beautiful are his descriptions of the afterlife and the oneness of everything. He writes:

"Seeing and hearing were not separate in this place where I now was. I could hear the visual beauty of the silvery bodies of those scintillating beings above, and I could see the surging, joyful perfection of what they sang. It seemed that you could not look at or listen to anything in this world without becoming a part of it – without joining with it in some mysterious way.

Again from my present perspective, I would suggest that you couldn't look at anything in that world at all, for the word 'at' itself implies a separation that did not exist there. Everything was distinct, yet everything was also a part of everything else, like the rich and intermingled designs on a Persian carpet.....or a butterfly's wing."

For a man who beforehand was so certain in his ideas about the nature of consciousness to have an experience so powerful that it transcended the very essence of his beliefs is both as wonderful as it is astonishing. As his experience clearly reveals - we are conscious *in spite of our brain* and that in truth, *consciousness is at the very heart of existence.*

Auric fields, chakras and meridians

Our connections to other frequencies whilst we are restricted to decoding reality at the physical level are further enhanced through additional energetic mechanisms that are worthy of mention. These are the electromagnetic fields surrounding and interpenetrating us that are known collectively as the 'aura', the system known as the 'chakras' and the powerful circuitry that acts as 'the motherboard' of the physical body, the 'meridians'. Each of these arrangements helps facilitate the flow of information between the vibrationary energy levels of which we are composed, ensuring a healthy balance is maintained.

When considering the word energy, we must also think 'information' and at each successive level, continuous energy exchanges of information are taking place through the auric field obtaining to that particular frequency. In his book *The Universe of Silver Birch,* author and researcher Frank Newman writes "the manner in which these 'bodies' can be identified as separate units existing within close proximity is by each 'body' functioning on different frequencies, vibrations or wavelengths. Thus, we have within the aura an energy field identified by science as being electromagnetic in nature and 'functioning in the ultra-violet, visible and infrared ranges. It can therefore be reasonably assumed that 'multiple bodies' at varying electromagnetic ranges exist within the auric field, which confirms the teachings received." Newman further asserts that, "one field cannot exist within another field without interaction taking place and producing results of great significance". This simply means that any exchange taking place at one level or frequency is reflected in similar terms within the related fields. Newman obviously had a grasp of the true nature of life as later in the same book he refers the 'holographic nature of the auric field' with 'every fragment possessing the essential characteristics of the complete field'. He also quotes the findings of Professor H.R Burr who states that the auric fields are "responsible for the maintenance of the human body by controlling the rebuilding and replacement of molecules and cells in the precise pattern of the original". From this Newman concludes, "We can therefore accept that they are stating that the DNA control of the physical form of a body stems from the electromagnetic nature of the auric field surrounding it." Sound familiar?

The term 'chakra' is an ancient Sanskrit word that means 'wheel of light'. This has become popular within 'new age' circles and great emphasis is placed upon being able to 'open' or close' what are essentially spinning vortices of energy connecting us to other levels of awareness. There are essentially seven main chakras situated along the main spinal column and multiple smaller ones located throughout the body and for ease; these are often numbered *(Fig 18)*.

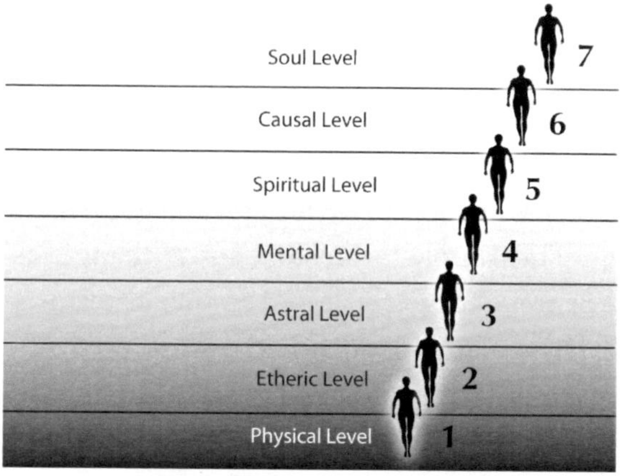

Fig 18. The chakras within the auric fields

So, working from the top down, the major chakras in descending order are, (7) the crown chakra, situated on top of the head - this is our true 'connection' with the interconnected universe and our 'antenna/receiver'. (6) The 'third eye' chakra situated in the brow and connected to both the pineal and pituitary glands in the brain (both of which are attributed to 'psychic sight' and provide a balance of intuition and intellect). (5) The throat chakra, linked to creativity and communication. (4) The heart chakra, located in the centre of the chest and responsible for connecting us to higher levels of communication. This is the centre of love and nurturing. (3) The solar plexus chakra, situated in the stomach area and connected to feelings and emotions (ever felt as though you've been kicked in the stomach by a hurtful remark)?

This centre is also connected to social values and issues of self-esteem. (2) The sacral chakra, found just beneath the navel closely linked with sexuality and reproduction. It is also linked to issues of financial security and relationships. (1) The base chakra, located at the foot of the spine and said to be the centre for *'kundalini'* energy (the coiled serpent) that when awakened surges through the central nervous system, eventually pouring through the top of the head and connecting us with higher levels of being. The kundalini/chakra system is also electrical in nature and forms part of our connection with the electrical aspect of the earth and the universe itself.

Each chakra point is associated with a different frequency and colour, rather like the colours in a rainbow and these are often reflected in photographs taken of the auric field. The colours are as follows:

1) **Red** – Root
2) **Orange** – Sacral
3) **Yellow** – Solar Plexus
4) **Green** – Heart
5) **Turquoise** – Throat
6) **Violet** – Third Eye
7) **White** – Crown

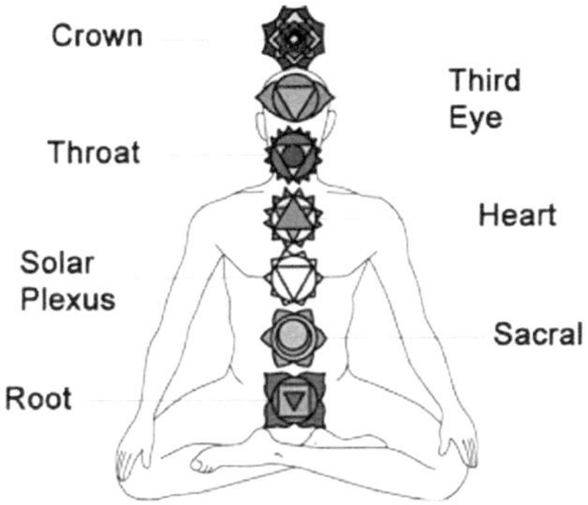

What is also highly significant about the chakra system is that it is fundamentally linked to the way in which we associate through our physical form throughout life. The base chakra for instance, forms the primary focus of our early development from birth until around the age of three when we are largely concerned with establishing ourselves within the physical world, whilst the sacral chakra becomes more dominant from the age of three until seven as we are forming and developing relationships. It is also at this early age that our *conscience* begins to emerge. From seven through to the age of twelve and the onset of puberty our third or solar plexus chakra is more influential whilst the heart chakra fully emerges in our teen years. This fourth centre is in one sense the meeting place between the lower and higher levels of being which is why the heart area is such a powerful energy point within the body. The fifth chakra, located in the throat area is also significant in the years leading up to early adulthood in terms of the ability of an individual to communicate and establish an identity in the world and the brow chakra also emerges fully during the years between seventeen and twenty-one. A fully mature and awakened adult will also have a crown chakra that is fully open and functioning, enabling a true connection to higher frequencies and non-physical dimensions.

A word of caution here; the interest in both chakras and aura photography has grown immensely over recent years and the advent of 'aura cameras' has allowed images of the human energy field to be more readily available to larger number of people. However, many of these cameras are inferior in that they offer computer enhanced images which although entertaining are not necessarily scientifically accurate. The process commonly known as *Kirlian photography*, pioneered by Semyon Kirlian (1898 – 1978) long before the advent of computer technology, offered more reliable images with astounding detail and clarity. A Ukranian engineer, Kirlian and his wife Valentina found that when subjects were placed within a high voltage electrical field a luminous aura appeared around them. Subsequent investigations showed a relationship between the images captured on film and some physical and psychological states within the subject. A typical example is shown in *(Fig 19.)* which reveals a degenerative lower back problem typified by 'heavy' feet and the absence of energy around the little toe (on both feet). A competent therapist taking such a reading would know immediately that a problem existed because of the 'gaps' in the energy field and the discrepancies around the feet, being able to offer suitable treatment as a result. There are also numerous images in existence depicting energy fields around humans, animals, plant life and inanimate objects, many showing the most beautiful and unique patterns of light and colour.

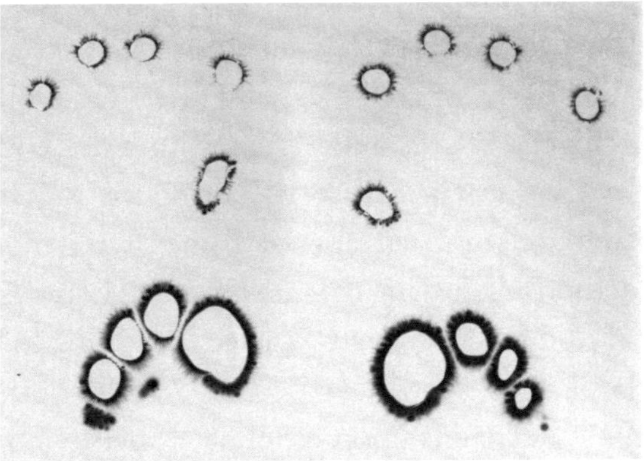

Fig 19. An aura photograph of hands and feet (chronic back problem)

Ten years prior to Kirlian's discovery a Russian embryologist, Alexander Gurwitsch (1874-1954) who along with Leonid Mandelstam (1879-1944) helped to develop the formulation of the Morphogenetic field theory showed that living tissue released photo-emissions in the ultra-violet range of the spectrum that showed significant correlations between biological and physiological functions. Gurwitsch named the phenomenon *mitogenetic radiation* as he believed that this light radiation allowed the morphogenetic field to control embryonic development.

More recently further breakthroughs have seen the development of the biophoton theory. Biophotons, or ultraweak photon (light) emissions from biological systems are weak electromagnetic waves in the optical range of the spectrum. Put simply, light is being emitted from the living cells of plants, animals and human beings. These emissions cannot be seen by the naked eye but can be measured using specialist equipment. According to biophoton theory, the light is stored in the cells of organisms or more precisely within their DNA molecules and when released or absorbed creates a dynamic web of light connecting all cells, tissues and organs within the body, vital to all life processes. Furthermore the discovery of biophoton emissions adds weight in support of some of the more unconventional methods of healing that are based on homeostasis (self regulation) including homeopathy and acupuncture. In his widely acclaimed book *Biophotons – The Light in Our Cells* author Marco Bischof (1947 -) expands upon biophoton theory in far greater detail with information on the holographic field of the brain and nervous system, suggesting that within the whole organism it may form the basis of memory and other phenomena associated with consciousness.

The term 'meridian' is one that acupuncturists will be familiar with and the passage of so-called *'chi'* energy around the meridian system is another way that information is distributed. When viewed using special imaging technology this system resembles a computer motherboard comprising a series of junctions linked together by interconnecting routes allowing energy to move freely around the body. Occasionally there are blockages within this network and it either slows down or in some cases speeds up, often resulting in an imbalance that eventually manifests as a health related problem. The meridian network forms part of the same decoding mechanism that connects us with other levels and dimensions. A skilled practitioner is not only able to detect and pinpoint the source of a blockage but also release it by applying special needles or a very mild electrical current at specific points. Often the area in which pain or discomfort occurs may require treatment at a different location upon the body but relief is felt because of the connection between the two points allowing the chi to begin flowing normally again. A very simple analogy might be to consider a blocked drain, resulting in an overflowing sink. The immediate cause of concern might be an excess of water on the bathroom floor, but until the source of the problem is located and treated the problem remains. The skill is in knowing where the blockage may lie.

Because we are at one level holographic in nature, it comes as no surprise that healing techniques based upon knowledge of the channels and zones representative of the major organs and areas of the body are so powerful. We have already mentioned acupuncture but others include acupressure, reflexology, iridology and emotional-freedom-technique (EFT). Related to the same holographic model we also find, naturopathy, homeopathy, holographic memory release (HMR) and bio-resonance therapy to name but a few. New discoveries and therapeutic interventions are springing up on a frequent basis as our knowledge rapidly expands, transforming the face of modern medicine.

The Earth too, has its own meridian system through which electrical and vibrational energy flows – commonly known as *ley lines.* These are often dismissed as being simply geographical alignments between places of historical importance such as ancient monuments and megaliths but they are much more than this. The ley line system exists as an *independent circuitry with the capacity to affect consciousness* and ley lines are part of the earth's energy system and the ancients knew this. Their stone circles, temples and other structures served to commemorate and perhaps in some way access this energy. It is no coincidence that stones containing crystalline structures were used because by their very nature they help to transmit and amplify energetic information in much the same way that our own DNA does.

Once more we discover a connection to light because ley lines are associated with cosmic forces originating outside of the earth. The lines are energetic in nature penetrating our planet and extending above it vertically, often passing through buildings. Like water lines, ley lines form a global network of force fields that can both affect human behaviour as well as being influenced by it, so here too it seems that consciousness in its varying degrees plays an interactive role. Consciousness, as we are discovering at every turn, is the key ingredient to life itself.

One journey, many lives

As we have established, we are multidimensional beings functioning simultaneously upon countless frequencies of existence and for ease of understanding we have catagorised these in the form of seven planes with seven corresponding bodies or vehicles of expression. Our individual consciousness, being expansive and not confined to any one level, is connected to *all* levels but at any one time its focus of attention is for the most part dictated by the lowest denominator or frequency at which it expresses. If we draw upon the example of a camera, then we will appreciate that although the camera lens may be able to take in the totality of that towards which it is pointed, it is that upon which the lens is *focused* that defines the clarity of the picture. Thus, here within the physical realm our main focus is essentially on decoding this frequency. However, when our consciousness departs at 'death' (refocuses on another level) our attention shifts and the astral plane becomes our new reality until such time that our vibrations either quicken to another higher plane or we return to the physical level once more. This 'back and forth' concept is better known as *reincarnation* or *the transmigration of the soul.* Although for many the jury is still out as to whether true or not, with valid arguments from all sides, the concept of reincarnation simply cannot be ignored and it is here that we go next.

Summary of main points:

- We exist simultaneously upon many levels or frequencies
- As we develop and our consciousness expands we become increasingly aware of these other dimensions
- Our individual consciousness is not confined to any one level
- We have a vehicle or body through which to express upon each level
- There are electromagnetic energy fields, collectively known as the Aura surrounding each body
- We have other connective conduits that allow the flow of energy information to and from higher levels. These include the chakra and meridian systems that are mirrored in some ways by the Earth's own energy grid
- Our individual consciousness, although expansive and able to shift between frequencies, remains largely focused upon the lowest plane of development to which it has progressed

Chapter Four

Lives Within Lives

"It is the secret of the world that all things subsist and do not die, but only retire a little from sight and afterwards return again. Nothing is dead; men feign themselves dead and endure mock funerals and there they stand looking out of the window, sound and well, in some strange new disguise."

Ralph Waldo Emerson (1803-1882)

If for many people, the idea that they exist upon multiple frequencies is difficult to accept then the notion that they also at some point return to the 'physical' level of being can be even harder to comprehend. Yet this need not be so if there exists an understanding of the ways in which consciousness operates. Part of the problem of acceptance may lie with the terminology being used. Reincarnation is an old term with religious connotations and does not sit particularly comfortably alongside or within the quantum world. But change it for a more scientific sounding term such as metempsychosis or transmigration and it starts to look a little more plausible and 'user friendly'.

Once we recognise that our non-physical self uses vehicles of expression (resembling bodies) that coexist side by side and are interpenetrating of each other and that once these have served their usefulness over a certain period (a 'physical' lifetime) we withdraw (re-focus) to other frequencies, it makes perfect sense to consider that at some point we may wish to recreate new forms once more. Perhaps we have unfinished business at a particular level – things we failed to address or could not complete the first time around. Or maybe we desire to help a close friend or loved one fulfill a part of their spiritual journey. Alternatively it could be that we need to learn a specific life lesson or experience a particular effect such as pain or poverty as a way of opening up our sensitivity. Either way, the opportunity exists within the parameters of our own making.

As with all aspects of natural phenomena, there are cycles which govern the movements between frequencies ('life and death') and all that exists in between and it should never be thought that any major shift from one dimension to another is anything other than in accordance with the fundamental working of natural law. Events which seem random, even at the quantum level where 'cause and effect' appears suspended, are due to the outworking of an *implicit order*. We may not always recognise this principle in operation and are seldom consciously aware of cycles of great magnitude but we reap their consequences nevertheless.

In regards to the number of incarnations we experience and the length of time spent in the non-physical dimensions between 'lifetimes' it seems that there are no set periods from which we can determine exact measurements. Just as we each experience vastly different lengths of life span, with some living beyond a century and others but a few moments, so it seems too that the time spent between incarnations varies enormously. This variable span does though, seem to be tied in with the lives of others with whom we share personal relationships and it would appear that for the most part, the opportunity to reincarnate is not available until all but the closest of these earthly ties and their cycles have completed. In simple terms, we cannot or do not reincarnate whilst those with whom we have close links are still in the midst of their own earthly sojourn. A child for example, would not under normal circumstances come back into another body whilst its former parents were still alive on earth. Once all the cycles have been outworked across several generations only then can those who had incarnated earliest begin to consider their options to do so again. This cycle, which I refer to as the *family-soul-cycle* will typically encompass a time span of around one hundred to two hundred years, allowing for a normal current life term of between seventy and one hundred years for each individual. As *(Fig 20.)* illustrates those who are the great, great, great, great grandparents (level 1) of today's infants (level 6) are still unlikely to reincarnate because even though there is no direct link between them and the current generation (other than genetically) their own children (level 2) and certainly their grandchildren (level 3) are likely to still be alive on earth.

The explanation for this is both logical and beautifully crafted because where there are strong, loving ties of consciousness that span generations it seems highly probable that an individual leaving this frequency at 'death' would expect to meet up once more with their former family members and close acquaintances. If the creator is indeed a 'loving God' of supreme intelligence why would life be so designed as to further separate individuals after death by allowing those who had passed earlier to reincarnate?

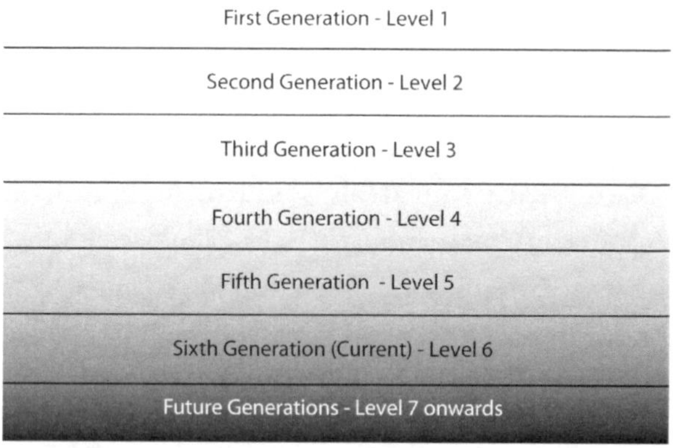

Fig 20. Generational ties within the 'Family-soul-cycle'

There are of course exceptions to every rule and there exist examples of souls who have apparently reincarnated relatively soon after the death of their former 'self'. Within certain races, most notably the Indian culture, the idea of reincarnation appears to be more widely accepted and there are some recorded cases of this kind. In 2001 *The Journal of Scientific Exploration* published a paper *(Vol. 15, No. 2, pp. 211–221, 2001)* by Satwant K. Pasricha from The Department of Clinical Psychology, National Institute of Mental Health and Neurosciences in Bangalore covering the investigation of cases of reincarnation in Southern India, which highlights the question 'Why so few reports?', stating that of the 450 cases professionally investigated in the country, all but one involved children from the North of the country. Suggesting that cases from the South may have been under-reported, the document then continues to detail several occurrences from that region and it makes fascinating reading.

Take the case of a subject named Anuradha for instance, of whom Pasricha reports:

> *"Anuradha (of Karnataka), who was born into a Hindu family, remembered the life of a Moslem who lived in Kashmir, where apples are grown in abundance. She was fond of apples from a very young age and would pick up only apples out of the many fruits offered to her. She also insisted on eating mutton every day, which her family provided. As this case remained unsolved, I can not say for sure that it derived from her previous life, but it is congruous with her claimed life of a Moslem. Her parents, however, ate chicken, which Anuradha never liked."*

Pasricha continues with more startling revelations:

> *"As mentioned above, Anuradha, whose family members were Hindus, remembered years she used to assume the posture of saying Namaz (Moslem prayers), and if taken near the Hindu shrine at home, she would blow out the lamp, showing her dislike for the Hindu way of worship. When Anuradha started to write, she would write from right to left, as is the convention with the Arabic and Persian scripts. Anuradha also had a phobia of water from the age of 2 years on. She remembered having drowned in her (unverified) previous life. Her mother had to force her to take a bath. The phobia continued until Anuradha was 9 years old. Her mother wished her to learn to swim, but she refused to do this. Phobia is another common feature among North Indian cases."*

More recently Dr Michael Newton, PhD., a counseling psychologist and master hypnotherapist has written several best selling books detailing his own investigations into reincarnation and life between lives. These have been translated into over twenty-five languages and in 2001 the Independent Publishers Association at their annual 'Book Exposition of America' meeting awarded 'Metaphysical Book of the Year' to his second book. He is also the founder of 'The Newton Institute for Life Between Lives Hypnotherapy'. Dr Newton has documented many sessions in which people have, whilst in an altered state induced through the use of hypnosis and age regression techniques revealed information about their past lives. Many of his subjects also presented with current life issues that were later proven to be linked to past life events and experiences.

Newton recalls how he 'stumbled across a discovery of enormous proportions' whilst working as a therapist and found that he could actually see into the spirit world 'through the minds-eye' of the hypnotised subject. As his case files grew he began to notice similarities between clients descriptions of the afterlife or 'between life' states and he developed techniques that involved asking hypnotised subjects questions about the spirit world in a proper sequence to both eliminate intervention or corruption from their own subconscious mind (some people have been known to pollute or interfere with the therapy and fabricate memories either consciously or unconsciously) and to also unlock deeper levels of information. As Newton himself says 'One must have the right set of keys for specific doors'.

Authors Bruce and Andrea Leininger also present a very

compelling case in their book *Soul Survivor – the Reincarnation of a WWII Fighter Pilot.* Over sixty years ago, a 21-year-old Navy fighter pilot on a mission over the Pacific was shot down by Japanese fire. Many years later the authors' son James Leininger began revealing what seemed to be personal information about James Huston Jr., the pilot in question. James would dream about flying planes and his mother often had to wake him from his nightmares. When she inquired what he had been dreaming about he would say, "Airplane crash on fire, little man can't get out." In one video clip of James at the age of three, he casts his eyes over a plane as if he's doing a pre-flight check. On another occasion Andrea bought him a toy plane and pointed out what appeared to be a bomb on its underside. James quickly corrected her, insisting it was a 'drop tank', much to the amazement of his Mother. "I'd never heard of a drop tank," she said. "I didn't know what a drop tank was." Across the years James revealed much more detailed information about his former life, saying that the Japanese had hit his plane and that he had crashed. He remembered flying a Corsair and commented how they 'used to get flat tyres all the time.' He also told his parents the name of the vessel from which his plane took off – 'Natoma' – as well as mentioning the name of a man he flew with - 'Jack Larson'. After researching what his son had said, Bruce discovered that both were real. The Natoma Bay was a small aircraft carrier operating in the Pacific and Larson lived in Arkansas. A great deal of other information was forthcoming and Bruce believes his son had indeed lived a past life in which he was James M. Huston Jr., saying, "He came back because he wasn't finished with something."

The Leiningers also wrote a letter to Huston's sister Anne Barron, about their little boy and she too believes the authenticity of James' story, saying, "The child was so convincing in coming up with all the things that there is no way in the world he could know."

Although some researchers have cast doubts upon these and other instances where the factual evidence is overwhelming, there is little doubt that information from 'the past' is being accessed in a non-linear way by those who currently reside in this physical dimension. Whilst it is possible that a non-local resonant field could exist between discarnate and incarnate individuals, allowing information to be transferred across dimensions, or even that some residual memories from a former life could be retained at the quantum level, it is more likely in my opinion that the concept of reincarnation is true. I base this judgement on my knowledge and understanding of the journey of the soul and I firmly believe that the entire concept of birth – between life – rebirth, is both logical and in every way essential to the progress of each individual.

Following on from this, we can begin to formulate an image of what our spiritual pathway might look like. Utilising our familiar model of the various planes *(Fig 21.)* illustrates the progress consciousness takes as it slowly evolves through the lower, more dense frequencies to the higher, more refined states.

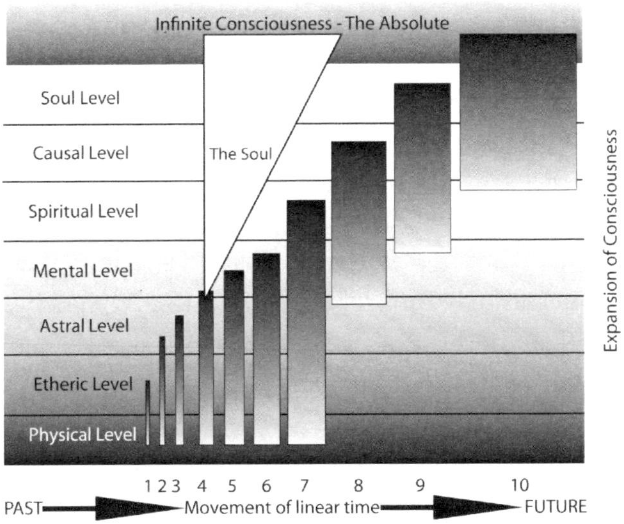

Fig. 21 Movement from lower to higher states of being

The evolution of consciousness though form

There are a *hierarchy of realities* through which life itself experiences and the illustration shows that as consciousness expresses through form at the lower end of the evolutionary scale (1) its experience encompasses only the physical and etheric levels, with no awareness or development of the astral body. *Pure infinite consciousness* at these levels has not yet individualised and operates through simple forms at the very lowest end of matter, such as single electrons through to more complex forms like viruses. There may be many 'lives' within this frequency range before the opportunity presents itself to experience an expansion into more developed vehicles and this evolutionary track enables a hierarchy of more complex systems to develop through which consciousness can gain greater expression, what could be referred to as the *implicit order of evolution.*

Moving above the etheric plane and into the next dimension of being we see the emergence of the astral level. There is still no true individuality here as far as consciousness is concerned but more evolved forms and species appear allowing for an increasingly broader experience (2). At the lowest end of this level we find rocks and minerals and also elements such as water. More advanced minerals such as crystals also exist here and as we have already discussed, water contains a form of memory that responds to words and emotions whilst crystals themselves have many properties that allow energies to be channeled or magnified with great effect. At the higher end of this level exists the plant kingdom and one of the main differences that we observe in comparison to the mineral reality is one of growth and reproduction. The emergence of a 'sex life' in terms of being able to reproduce (even with the help of insects) is a step up from the previous level and here the quality of consciousness is higher as is the response to stimuli. Plants as we have discovered also respond to the energy of thoughts and emotions in positive or negative ways.

Going further up the chain we reach the animal reality (3) and here we find that the astral reality really begins to exert a more powerful influence through a broad range of emotions. Movement becomes a choice as limited freewill is experienced for the first time, as is interaction across species with animals communicating with one another and in some instances with humans.

As within all levels there exists a hierarchy of forms and it can clearly be seen that certain species are more advanced than others. Insects for example - another of the forms found at this level, are able to make limited choices and decisions about their movement and behaviour but quite clearly they are not as intelligent as groups higher up the evolutionary ladder whose options are greater (4). This is because the emergence of 'mind' and the development of the brain within higher animal species enables consciousness to not only 'think' but more importantly *reason,* allowing for rational choices to be made that involve more than just instinctive behaviour. Whist some less developed creatures at the lower end of this range may operate through a group consciousness or be connected through operation of the aforementioned *morphic field*, the likes of humans and some of the more evolved species such as dolphins, whales, cats, dogs, apes, horses and others clearly operate more independently of each other even though connected at the most fundamental levels.

The individualised soul

We should mention at this stage the emergence of the soul. Until it reaches the astral levels of its journey, consciousness does not fully individualise and should be considered as being part of a group dynamic. All insects, birds, fish and animals - except for those whose vibratory rates have been quickened through a mechanism that has allowed for a more expansive expression of consciousness, exist within a group soul species. Only when consciousness is able to individualise can the true meaning of the term 'soul' be employed. In some lesser creatures this does happen, particularly where a loving relationship with a more evolved species such as a human (yes, we are apparently more advanced) occurs, enabling a unique expression to develop. In the case of us humans, each of us has a soul that, even though still part of a group situation at the higher levels, is nevertheless unique and distinct, a quality that is forever retained once established. The fact that beloved pets as well as humans are able to return through mediumship to prove their continued existence after death is proof that their unique identity survives beyond the physical level. If this were not the case and all animals returned to a group soul status at the end of their physical incarnation it would be impossible for them to impart verifiable information to their former owners and numerous cases have been recorded featuring evidence of all types, including photographic, auditory and kinesthetic amongst others.

As individual consciousness continues to expand through successive incarnations spent within linear time – which may involve hundreds or even thousands of individual lives (5,6,7 - *Fig 21.)* so there eventually comes a point where the experiences of life within the lower frequency ranges have served their purpose and there is no longer any need to reincarnate. From here on, life continues unabated upon the mental planes (8 - *Fig 21).* Whilst the astral shell or former astral body may be retained by leaving a portion of consciousness within it (rather like leaving an old computer still connected up to the mains, ready to use if required) the expanded soul is freed to explore and experience these higher, more refined non-physical frequencies as the journey continues.

Darkness and light

I should point out that there are lower, denser and darker non-physical dimensions in addition to the higher planes and it does not always follow that a soul will consistently follow an upward spiral. But if during a physical lifespan excessive negative *karma* has been built up and retrograde steps have been taken, a spell within these lower realms followed by reincarnation to the 'physical' level will be sufficient to enable the imbalance to be corrected. This is not by the way, any kind of punishment meted out by a vengeful deity. Rather it is as a direct consequence of an individual's own actions and the fulfillment of natural law. Once all physical lives and experiences have been exhausted and all lessons learned, then and only then can a continuing upward progress be pursued.

The natural question to ask is 'What is life like within each of these realms?' If we consider what we have already discussed regarding the nature of reality and the ways in which we both decode information and manifest that reality, it seems logical to assume that each dimensional plane is a creation of the collective consciousness residing there – and we would be right. As correspondingly low as each soul can sink or as relatively high as it can ascend, so it experiences the worlds of its own making. These realities are the living constructs of consciousness in direct accordance with the nature of each soul. Individually and collectively the vital energies combine to reflect the nature of all who dwell within each frequency. There can be no aberration to the operation of the law underpinning this principle and no escape for the impoverished spirit who comes face to face with the consequences of its own actions. This is perhaps the purest example of karma in action for not only does the current mind-set of each person have a bearing upon their experience but more assuredly, the very nature, character and measure of spiritual development in direct proportion to every deed done upon earth plays a part in determining the place in which a soul finds itself after leaving the 'physical 'realm.

To explore the nature of each sub-division of the non-physical dimensions would require a book devoted to the subject, such is the vastness of the topic. Fortunately there are a number of writings devoted to the nature of life after death worthy of consideration. Perhaps the most famous of these is *Life in the World Unseen* by Anthony Borgia. The contents were communicated from the spirit world it is said, by Monsignor Robert Hugh Benson (1871-1914), a son of Edward White Benson, former Archbishop of Canterbury following his death in 1914. Benson uses very descriptive language to convey his message and the imagery conjured up by his words reveals worlds of contrasting beauty and ugliness. Speaking of a visit to the higher planes he says:

"We were in a dominion of unparalleled beauty. There is no imagination upon the Earth-plane that can visualise such inexpressible beauty, and I can only give you some meager details of what we saw in the limited terms of the Earth-plane."

"Stretching before us was the wide stream of a river, looking calm, peaceful, and overwhelmingly lovely as the heavenly sun touched every tiny

wave with a myriad tints and tones. Occupying a central position in the view, and upon the right bank of the river, was a spacious terrace built to the water's edge. It seemed to be composed of the most delicate alabaster. A broad flight of steps led up to the most magnificent building that the mind could ever contemplate."

"The whole edifice was exclusively composed of sapphire, diamond, and topaz, or at least, their celestial equivalent. These three precious stones constituted the crystalline embodiment of the three colors blue, white and gold, and they corresponded with the colours which we had seen before in the robe of our celestial visitor as we had seen him in the temple, and which he carried in such an immense degree. The blue, white and gold of the jewelled palace, touched by the pure rays of the great central sun, were intensified and magnified a thousand fold, and flashed forth in every direction their beams of the purest light. Indeed, the whole edifice presented to our bewildered gaze one vast volume of sparkling irradiation. We at once thought of Earthly topaz and sapphire and diamond, and we pondered how small stones of purity were only tiny objects that could be held between the forefinger and thumb. And here was an immense glittering mansion entirely built of these precious stones, and of such stones that the incarnate have never beheld - nor are they ever likely to behold while they are incarnate."

Contrast that to his description of a visit to the darker levels:

"At close view it became clear that these dwellings were nothing more than mere hovels. They were distressing to gaze upon, but it was infinitely more distressing to contemplate that these were the fruits of men's lives upon Earth. We did not enter any of the shacks - it was repulsive

enough outside, and we could have served no useful purpose at present by going in. Edwin therefore gave us a few details instead."

"Some of the inhabitants, he said, had lived here, or hereabouts, year after year - as time is reckoned upon Earth. They themselves had no sense of time, and their existence had been one interminable continuity of darkness through no one's fault but their own. Many had been the good souls who had penetrated into these Stygian realms to try to effect a rescue out of the darkness. Some had been successful; others had not. Success depends not so much upon the rescuer as upon the rescued. If the latter shows no glimmer of light in his mind, no desire to take a step forward on the Spiritual road, then nothing, literally nothing, can be done. The urge must come from within the fallen soul himself. And how low some of them had fallen! Never must it be supposed that those who, in the Earth's judgment, had failed spiritually, are fallen low. Many such have not failed at all, but are, in point of fact, worthy souls whose fine reward awaits them here. But on the other hand, there are those whose Earthly lives have been spiritually hideous though outwardly sublime; whose religious profession designated by a Roman collar, has been taken for granted as being synonymous with spirituality of soul. Such people have been mocking God throughout their sanctimonious lives on Earth where they lived with an empty show of holiness and goodness. Here they stand revealed for what they are. But the God they have mocked for so long does not punish. They punish themselves."

"The people living within these hovels that we were passing were not necessarily those who upon Earth had committed some crime in the eyes of the Earth people. There were many people who, without doing any harm had never, never done any good to a single mortal upon Earth. People who had

lived entirely unto themselves, without a thought for others. Such souls constantly harped upon the theme that they had done no harm to anyone. But they had harmed themselves."

The world famous spiritual healer and medium Harry Edwards (1893-1976) also published a book in 1976 called *Life in Spirit* that is comparable to Borgia's in its narrative of the ethereal realms. Although apparently written from the author's own personal knowledge, the book features a number of similar themes and its descriptive passages are analogous in many subtle ways. Whilst it is possible that Edwards was influenced by the earlier publication it seems more likely that he was himself inspired from a higher intelligence, just as Borgia was.

Spiritual expansion

The idea of success in material terms is often measured against a person's social standing, career success, life achievements or wealth. Yet spiritually, these count for very little and what is important is the way that we have lived our lives, the people we have helped, the service we have rendered and the good we have done. These aspects not only help to develop our character but more importantly add to the quality of our soul. Is it any wonder then, that where we find ourselves after departing the physical world is in direct proportion to what we truly are? It can be no other way, because the very essence of our being must always resonate with the field into which we involuntarily place ourselves. We cannot be who we are not and we cannot exist upon a plane with which we do not harmonise. As human antennas we are constantly transmitting and receiving the signals of life that attract us to places, events, people, life situations and environments where we are meant to be at any given moment. In the life beyond the physical universe we emerge into the exact time and space that we have generated for ourselves by all that has gone before and all that we have become. Because we are living energy, we have our own unique signal that identifies us as an individualised aspect of infinite consciousness. We could think of this as a kind of national insurance number or PIN code but on a universal scale. There is no one and nothing quite like us - true individuality, existing within and encompassed by the whole. Alone yet together, apart yet connected.

As our unique consciousness unfolds so does our ability to connect with other forms of creation and with the very source of life itself. Just as consciousness is restricted through its expression at the lower physical levels and is largely unable to affect its environment in any meaningful way, so as it ascends to higher levels the ability to influence and interact with natural law increases. This process may begin, even whilst incarnated upon earth but most assuredly develops with the procession through to higher dimensions. Having stepped out of the seemingly endless carousel that is reincarnation and ascended the ladder of spiritual progress (9,10 *Fig 21.*) a soul will find that its ability to transcend many of the restrictions it experienced whilst on earth will have endowed it with the most wonderful powers. It will most certainly be aware of its past lives and of the information retained at the causal level as well as seeing more clearly the truth that it had previously been denied. It will become the embodiment of universal love, understanding its connection to the absolute and the oneness of all creation. The requirement for an energy body depicting the recognisable male or female physical form will eventually be dispensed with as the essence of pure being emerges and the all-encompassing white light of wisdom and purity radiates from within. At this point there can be no return to the lower frequencies and even to connect directly with those existing upon the densest levels is quite impossible without the help of intermediary souls who act as a conduit through which communications can flow. Within such lofty spheres we can but imagine the infinite beauty of life and our language is totally inadequate to convey such majesty. Yet this is what awaits us all.

Energy curves and synchronicity

Returning to our current plane for a while, have you ever noticed how events occurring in your life often seem to have a synchronicity about them, as if there was some kind of process happening that you didn't quite understand but just happened to work in your favour? You know the kind of thing; a car pulls out right when you need a parking place or some money turns up unexpectedly just as a large bill drops through your letterbox. These kind of happenings could be dismissed as coincidence but when you understand the way in which energy operates at the quantum level you will know exactly how and why they happen, especially when you recognise that your own spiritual awakening and expansion is helping to facilitate this very phenomenon.

Consciousness is very powerful and infinite consciousness is all-powerful. The shift from ignorance to knowledge and from the lesser to the greater delivers its own rewards to the developing soul. Like finding treasure that has been hidden away, buried in the sand for centuries, our spiritual journey through the frequencies enables the inner power of the soul to be revealed. It is as if nature herself bows down before us, never in totality but always in degrees directly proportional to our spiritual essence. As we have seen, we each broadcast our own unique signal that both identifies and connects us in numerous ways to each other and to the very fabric of life. There exists an energy exchange mechanism that is both subtle, yet powerful and that coincides with our own emerging inner potential. What could be described as *energy curves* form the basis of our connectedness and transforming power with all that is around us. As our own consciousness expands and our inner force increases so are we able to more fully interact with, manipulate and directly influence events. Taken to the extreme, we reach a point where we have mastery over nature and the manifestation of form – we become Gods in miniature. The miraculous becomes attainable and is known to us.

Thus we can begin to see the effects of an expanded consciousness and an increased energy curve within the heightened states of some elevated individuals who have begun to demonstrate phenomena that can be classified as extra-ordinary or extra-sensory. Psychics and mediums, spiritual healers, intuitives, empaths, out-of-body travellers, remote viewers and those whose have gifts that science cannot comfortably explain all have extended energy curves. Clairvoyant vision is a classic example of how consciousness can operate outside of the physical dimension in an expanded state of awareness. The energy curve of the clairvoyant extends way beyond the etheric level and into the astral frequency. One who is classified as a seer typically has an energy curve that extends well above even the astral, perhaps into the spiritual or causal ranges. The world famous Indian guru and mystic Sri Sathya Sai Baba (1926-2011) who passed as recently as 2011 was it is claimed, able to manifest *Vibhuti* (sacred ash) and other small items such as rings, necklaces and watches apparently out of thin air. Although some claimed that these were mere conjuring tricks there are many who remain convinced that he was a genuine *Avatar*. Interestingly, Sai Baba claimed to be the reincarnation of Sai Baba of Shirdi (unknown – 1918) who was considered a spiritual saint and a miracle worker.

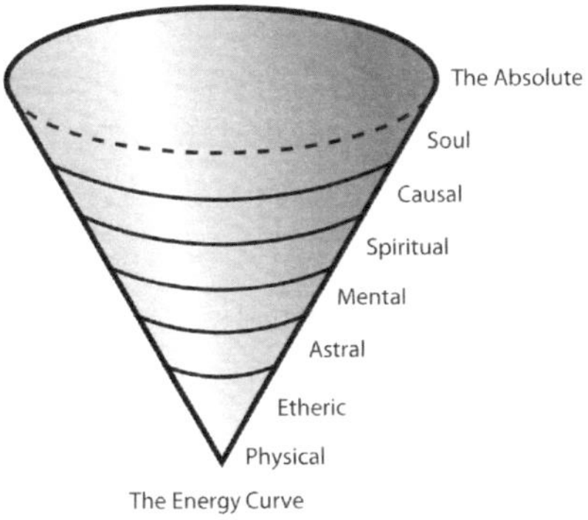

Fig. 22. The expanding energy curve

Edgar Cayce (1877-1945), the famous 'sleeping prophet' was another who demonstrated remarkable powers. Cayce was able to give detailed readings and diagnose many conditions of patients across a wide range of problems and during a period of 43 years from 1901 to 1944 gave over 14,000 such readings whilst in a hypnotic or 'sleeping state'. The 16th Century French seer Nostradamus (1503-1566) was also famed worldwide for his many far-reaching prophecies whilst more recently an upsurge in the popularity of mediumship has again produced many modern day prophets, some more accurate and reliable than others. I know from my own experience as a medium that insights into future events can and do occur and sometimes these can relate to happenings many years or even centuries hence. In my view these kind of prophetic visions cannot be explained by the morphic field theory or by resonance because they do not necessarily involve a connection between two or more individuals sharing the same quantum field. Rather they are examples of consciousness being able to extend its reach across and beyond normal *linear time,* entering into the realm of *global time* where past, present and future can be accessed more holistically or in a holographic way. I have also witnessed first hand on numerous occasions the power of simple prayer by one or more possessing the acquired ability to influence and affect the outcome of events in a positive way. In each instance these happenings were always commensurate with natural law and never contravened the freewill of any individual. Prayer you see, has nothing at all to do with religious belief or practice – but has everything to do with positive belief and intent, allied with the desire to help and be helped for all the right reasons. Those reasons always, without exception, encompass selflessness and altruism of the highest order. Why? - *because they have to.* The universe and everything within it is encoded that way and spiritual development brings its own rewards, every time. If consciousness expressing through less evolved forms is inhibited and restricted (because it has a lower energy curve) so the very same consciousness experiencing through a more evolved form must have an enhanced ability to connect and influence that with which it comes into direct contact in a benevolent way because its energy curve is vastly expanded and more vital. The great spiritual thinker and writer Ralph Waldo Trine (1866-1958) understood this great truth when he wrote:

> *"A miracle is nothing more or less than this. Anyone who has come into a knowledge of his true identity, of his oneness with the all-pervading wisdom and power, this makes it possible for laws higher than the ordinary mind knows of to be revealed to him."*

Thus it is, that as the layers of self-illusion peel away, we each become more self- aware, self-empowered and self-realised and our ability to sense and experience beyond the level at which we, for the most part temporarily function, expands. Hitherto unknown powers start to emerge and their impetus carries us forward into realms that were once beyond our reach, even though always there.

Group souls

Any discussion about reincarnation and the path of the soul invariably leads us to question what awaits us at the summit of our journey. Whilst it can be reasonably argued that an infinite journey offers infinite possibilities and that perfection is probably not attainable, the human mind, if not the consciousness that functions through it, desires structure. For most of us it is easier to grasp a concept or to understand a complex teaching if we can get a mental picture in our head. The tangible we can appreciate more easily – the abstract we have difficulty with. That is why this book contains many illustrations, some of them deliberately over-simplified, in an attempt to convey meaning to very complex subject matter. Yet the very nature of non-physical form and higher frequency dimensions takes us into areas where familiar reference points become blurred and the accepted norms are once more open to question. The quantum world with its extraordinary features and odd behaviour is one such notable arena where our ideas and views are challenged to the extreme and the ethereal planes are another. We have already glimpsed the higher planes of the spirit world and the ever-expanding state of individual consciousness and here upon approaching the zenith of human progress we encounter our own *soul groups.*

Each of us shares an affinity with others of a similar nature, whose own journeys have followed a similar, but never identical upward spiral. Magnetically we are drawn together like filings attracted by a powerful field and gravitate into a coherent whole. Never is our individuality compromised or our unique essence lost or diluted. Rather, we become the whole (group) and the whole (group) becomes us. *(Fig 23.)* details this hierarchical structure in a simplified way.

Every soul, having attained the heights of spiritual enlightenment and left the lower, denser fields far behind finds itself enveloped within a rarefied atmosphere where falsehood and illusion have been all but stripped away, revealing a pristine beauty within. Collectively each *sub-soul group* encompasses many hundreds or perhaps thousands of souls who share a deep affinity with each other, having journeyed through lives and dimensions to arrive at this point. Here there is no discord of any kind – just the blissful harmony of a collective whole in unison.

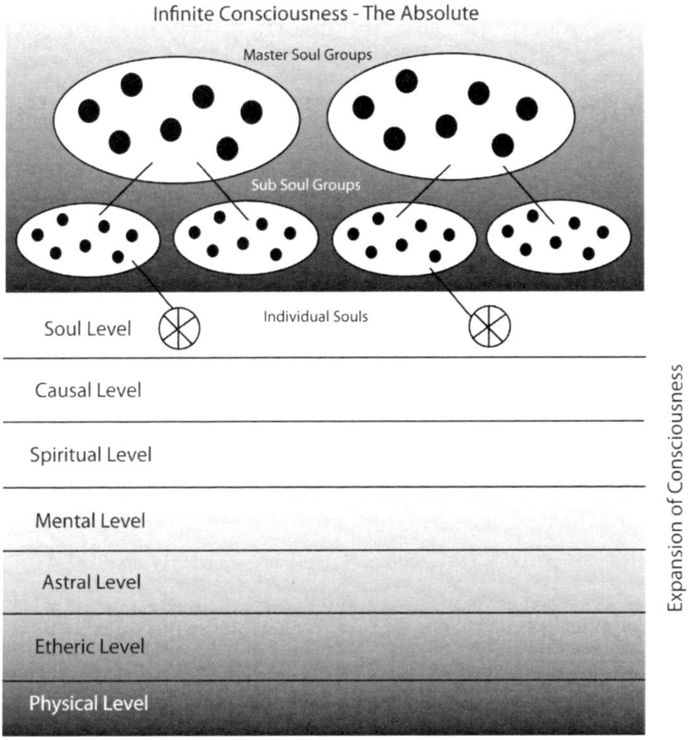

Fig. 23. The hierarchy of soul groups

The aspiration to reach even greater spiritual heights through the same implicit order that has always existed remains in-situ, as does the intrinsic desire to serve humanity and the opportunity to do this is freely available. Indeed, from these lofty spheres emerge many streams of wisdom teachings that pour down through the frequencies, eventually filtering through to the physical level to emerge either through mediumship or the inspirational works of teachers, philosophers, artists, musicians, healers and prophets. Just as the fruits of the tree are absorbed back into the earth to be recycled as the sap rises once more, so do the essential forces of those who have attained the heights of spiritual wisdom benefit all beneath them.

Above the level of sub-soul groups exist the *master soul groups* formed from the amalgamation of the former and existing in such a refined atmosphere that mere words are rendered meaningless when trying to describe such magnificence. We can but make vague attempts to imagine what life upon this plane consists of - such is the gulf between the pure consciousness of enlightened being and the heaviness of material mind and in speculating upon such a supreme ethereal existence we cannot assume anything. I would suggest however, that from this level we are able to partake in the creative process from which entire worlds, stars and galaxies are formed. So great would be the energy curve that we each possess, with an inherent ability to directly affect and interact with the quantum creative forces of life itself, that such a notion would be well within reach. Imagine the creative energy of such a soul collective to bring into being an entire universe of form – what a truly wonderful experience this would be. Perhaps such a soul group with the ever-present guiding light of infinite consciousness watching its every move created our own universe, with its myriad of life forms, natural laws and untold wonders?

The mystic within me ponders these concepts as perhaps

it also does within you, but with the passage of time we will surely know the answers to all that draws our enquiry.

Summary of main points:

• Life moves within cycles

• There is an 'implicit order' with all of creation

• A hierarchy of realities exists through which consciousness experiences

• Group consciousness and individualised consciousness reincarnate within successively more refined forms

• Expanding consciousness encompasses a greater energy curve

• An increased energy curve allows a more powerful exchange with life itself

• At the highest levels of spiritual attainment we exist within a soul-group

• Each soul-group is a collection of individual souls that share an affinity

• The divine essence within each master soul-group allows for infinite possibilities in the creation of form and the manifestation of worlds

Chapter Five

About Time

"The universe may be timeless, but if you imagine breaking it into pieces, some of the pieces can serve as clocks for the others. Time emerges from timelessness. We perceive time because we are, by our very nature, one of those pieces. "

Craig Callender (1968 -)

The nature of time has always fascinated us. Ever since the earliest cavemen observed the passage of the seasons, the movement of the heavens and the phenomenon of day and night we have wondered how the universe works and how the cosmological clock ticks. As with many of its observations, the mind formulates ideas and theories and as it accumulates facts these often become more complex and defined. The understanding of time is no exception and at one 'time' or another most of the world's great scientists or philosophers have speculated on what it is or might be. One of the problems we encounter though is that in life, we like constants. We feel secure in the knowledge that certain things are a certain way – day always follows night, apples fall to the ground, thunder follows lightning, summer comes along after spring and so on. But time is different to all of these as closer scrutiny reveals.

Let's look at some basic (and possibly disturbing) facts to begin with:

We may not actually live in the present – there exists a minute 'time delay' between what we decode in the brain and what we then experience as 'reality'. This is estimated to be around 80 milliseconds. In an experiment designed to prove this scientists asked volunteers to press a button, causing a light to flash with a short delay. After several attempts the volunteers began seeing the flash *immediately after pressing the button* – their brains had become used to the delay and had edited it out. The scientists being what they are (devious fellows) decided to remove the delay but to their amazement the volunteers then reported seeing the flash *before* they had pressed the button. Their brains, in attempting to reconstruct the events, actually *switched the order* resulting in them seeing the effect before the cause. David Eagleman (1971 -), a neuroscientist at the Baylor College of Medicine in Houston Texas and author of the book *Incognito: The secret lives of the brain* suggests that the perception of time is a *construction* of the brain and asks the question, 'How far in the past do you live?' Describing how when we snap our fingers, our eyes and ears register this information which is then processed by the brain, Eagleman tells us that by the time this occurs, the snap has already come and gone. Like a 'live' television programme that is beamed out with a few seconds delay (in case of any inappropriate language) our ability to process information always delays our conscious awareness of it.

If you live higher up you age faster – as ludicrous as this may seem it is true. Each time you climb the stairs you are ageing more quickly. Scientists have proved this by placing two atomic clocks, precisely synchronised on two tables. One table was raised by 33 centimetres and they discovered that the higher clock was running faster than the lower one at a rate of 90-billionths of a second in 79 years. This means in effect, that races living at higher altitudes age slightly faster than those living at sea level. Einstein's Theories of Relativity predicted that this would happen because gravity warps time as well as space and the closer we are to the ground the more that gravity slows down time. In addition to this extraordinary fact, time passes more slowly when we move faster – a phenomenon is known as *time-dilation.* The NIST (National Institute of Standards and Technology) ran further experiments focused on two scenarios also predicted by Einstein's theories using two clocks subjected to unequal gravitational forces due to their different elevations above the surface of the Earth. In the experiment the higher clock-experiencing a smaller gravitational force—ran faster.

The faster you travel the slower time moves - when an observer is moving, a stationary clock's tick appears to last longer, so a clock appears to run slow. Scientists refer to this as the *twin paradox*, in which a twin sibling travelling in a fast-moving rocket ship would return home younger than the other twin. The determining factor is the acceleration (speeding up and slowing down) of the travelling twin in making the round-trip journey. This was proven when one atomic clock was taken on a plane trip around the world whilst another stayed put at home. When the travelling clock returned from its trip – after some 50 hours and 800-kilometres it was missing around 230 nanoseconds. From this we can deduce that it *gained* time from being farther from the surface of the Earth, yet *lost* even more by travelling faster. Even stranger is that from the perspective of the clock on the aircraft, the clock left back home would have appeared to be running faster than normal.

We can each witness the same event at different speeds – Einstein, who seems to pop up time and time again (no pun intended) postulated that different people can witness the same event occurring at different speeds. In this scenario events that appear to happen simultaneously to people in motion may not appear so to someone standing still suggesting that we each have our own unique *timeline* that is *relative to our consciousness.*

This last point is I believe key to understanding that way that we each process time. For most of us humans, time appears to pass in familiar fashion – we usually wake up at the same time each morning, spend the day pursuing whatever activities we are compelled to undertake and eventually retire to sleep. Depending on where we live, our social status, health, age, occupation and circumstances our day is likely to be interspersed with the basic requirements of eating, drinking, mental and physical activity, excretion and sex (not necessarily in that order!). If we are fortunate enough to live in a place where the struggle to survive does not occupy all of our thoughts and energy we will also probably have sufficient leisure time to engage in pursuits such as reading, watching TV, playing sport, listening to music and so on. Where there are no timepieces to remind us of the exact hour or minute of the day our life will most likely be governed by the passage of the earth around the sun, giving periods of light and dark whilst life in a busy metropolis will inevitably revolve around the clock, regardless of whether it is night or day.

Man has always been intrigued by time and sought to discover and invent ways of measuring it to suit his needs. From the earliest sundials that measured the time of day by using the sun casting a shadow onto a cylindrical stone, through candle and water clocks, some of which were very sophisticated, and the advent of mechanical clocks around the 13th century, the development of time-measuring devices reveals a gradual movement towards more accurate devices. The advent of quartz crystal and atomic clocks has allowed the measurement of time to be recorded with great precision and the functioning of our modern world is entirely time-dependent.

Whilst the advent of accurate timepieces and the subsequent development of a time-managed global society has brought obvious advantages – we know what time our train home departs and when our favourite television programme is scheduled – we are also subject to an enormous degree of time-conditioning. Our lives seem to revolve around the clock and most of what we do centres around the measurement of time. There is however, a subtle difference in the precise measurement of time and our *perception* of time. Viewed from the perspective of the clock, the passage of time is always identical (give or take the odd nanosecond) whilst our *experience* of it may not be.

Most of us are familiar with 'waiting for a bus syndrome'. We find ourselves standing at a bus stop glancing at our watch every few minutes as we await the impending arrival of our transport. Time drags as the bus seems to take an eternity to appear and what may actually be only a few minutes feels like an hour. Conversely, we may discover to our disappointment that our enjoyment of a concert or dinner party is curtailed sooner than we would have liked because time has 'flown by'. Clearly, something profound is occurring here and it has to be linked to our internal perception. It is obvious that the way in which we decode the experience of time is not a constant. It changes from moment to moment as *we* change. The author of this is *consciousness*. Because we are decoding our experience of time in a linear fashion, it is subject to fluctuation and can never be fixed.

Consider for a moment the consciousness of a mayfly. The lifespan of an adult mayfly can vary from just thirty minutes to one day depending on the species. Yes, you read that correctly – just 30 minutes! Imagine living out your whole life in thirty minutes – it is difficult for us to conceive of such an experience. Yet, from the perspective of the mayfly the passage of time is vastly different to ours. For this humble insect, the experience of consciousness is acquiescent to the life purpose of its form. Here, consciousness has no need for the development of higher cognitive states that require a much longer life span – it can accomplish all that it needs within thirty minutes and from the perspective of self-awareness that length of time is experienced in the same degree that we would experience seventy or eighty of our years. Similarly many other short-lived species experience the passage of time relative to their life-purpose and evolutionary development.

Linear and global time

You may have figured out by now that time is not what we are conditioned to believe it is, so it is about time that I explained in greater detail my earlier reference to *linear time* and *global time. (Fig 24.)* illustrates how our experience of time shifts as consciousness moves beyond the limitations of its linear experience to an expanded awareness in which past, present and future combine into one whole that I call the *eternal now.*

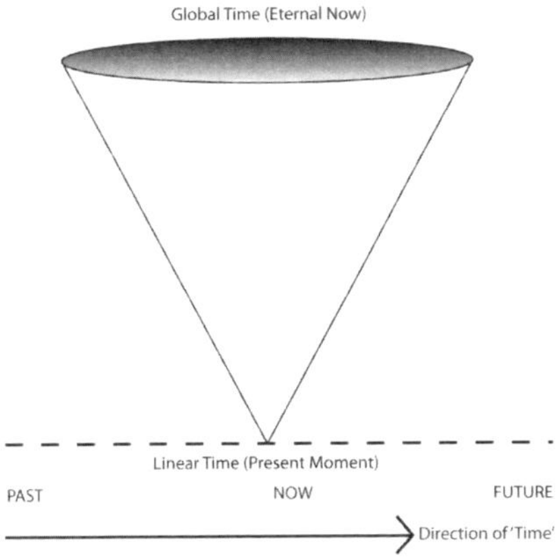

Fig 24. From linear time to the eternal now

If we consider our experience of time as an inverted cone we can clearly see that from the level of linear time experienced upon the physical plane, a definite direction emerges. This is known as the *arrow of time* and we each experience it sequentially. In other words, it is like watching a movie in which each successive frame appears before us as the story unfolds. The film cannot be reversed, it moves only forwards - although we may have memories of that which has passed. As the drawing shows, the tip of the cone permits only a limited experience of each moment, rather like the gramophone needle upon the vinyl record or the laser light upon the DVD disk. This though, is an illusion that is perpetuated by our inability to perceive the totality of time. There are reasons for this of course, because if at our present state of awareness we were able to experience simultaneously all that was, is and ever shall be, we would be unable to comprehend it.

So, at the physical level and to an extent upon the etheric and astral levels beyond, our consciousness decodes time in a linear fashion. Even here though, everything that occurs does so in *the now,* because in reality 'now' is all there ever is. Everything that happens can only do so within *this moment* – even our linear experiences. Eckhart Tolle (1948 -), author of the best selling book *The Power Of Now* highlighted this point when he wrote:

"Time isn't precious at all, because it is an illusion. What you perceive as precious is not time but the one point that is out of time: the Now. That is precious indeed. The more you are focused on time - past and future - the more you miss the Now, the most precious thing there is."

A word on Einstein's space-time

When Einstein developed his 'Theory of Special Relativity' he recognised that time and space are inseparable from each other – hence the term *space-time.* According to current thinking about the creation of the universe through what is commonly known as the *Big Bang* (more about this in the next chapter) a primal release of energy set the universe into motion. From the apparent onset of its creation around 14 billion years ago the universe has been continuously expanding and it appears that everything is moving away from one place at its centre, cooling as it does so. It is from the knowledge of this shift that we may discover a massive clue as to why time appears to move only in one direction and that is because *time IS the space in which it travels.* Whilst this phenomenon is always associated with the physical universe, I am suggesting that it still persists in the realms beyond physical matter, or at least those closest to it, because as our earth and its galaxy moves through space, so do the higher realms with which it is associated. It is only when we enter the higher frequencies beyond the material universe that our awareness of time is significantly different, yet just as all of the vibrational frequencies and planes that we have outlined to this point co-exist within consciousness, so too does time. Perhaps the recognised term space-time could be more accurately described as *space-time-consciousness* because this succinctly conveys the inseparable nature of this continuum.

It is perhaps the greatest paradox of all that the 'journey' through time and our 'expansion' of consciousness – indeed, the path of spiritual unfoldment itself, all occurs *in this moment.* Maybe this is why the great spiritual teachers tell us 'we can teach you nothing, we can only help you to remember what you have forgotten'. Viewed from one perspective, we are already enlightened and have journeyed through all that is – we simply don't remember.

A simpler way of looking at the complexity of this idea

might be to consider a household object to which we can more easily relate. Until the recent innovation of computer downloading, many of us held our entire music collection on CDs and having moved away from the outmoded idea of videotape, still have many of our favourite movies stored on DVD. There are also moves to develop holographic DVDs. Holographic discs are optical storage devices, similar to standard DVD discs that can store vast amounts of information. The aim of companies developing this technology is to produce discs that can store up to one terabyte (1024 gigabytes) of information. This process involves recording data in the form of three-dimensional patterns onto discs that are made of materials with high reflective indexes. Lasers are used to read the data from the disc, which in effect acts like a mirror. The difference between holographic discs and normal DVDs and Blu-Ray discs is that the information is recorded throughout the entire disk in layers, resulting in very high information density – perhaps this has a familiar ring by now? If we were to hold such an item in our hand we would recognise that an entire movie was encoded upon the layers of the disk and that even though it appeared as a blank, shiny flat surface to our naked eye, that would all would change once we placed it in our DVD player and pressed play.

The encoded information on a disk can be likened to *global time* in that everything is there in its entirety. Yet when the laser light decodes the information and we view the movie, it unfolds in a linear fashion. If it didn't then our brain would struggle to make sense of it and we would have great difficulty following the story.

Open and closed systems

There are of course many questions that arise from the concept of global time – in particular those involving time-travel and freewill but natural law seems to have all bases covered in respect to this. When considering the aforementioned *arrow of time* one of the most strikingly obvious features of the macroscopic world is that of *irreversibility*: we are born, age and then die (except that the 'real' us can't actually die) never the other way around. We can turn eggs into omelettes but cannot turn omelettes into eggs. A glass may fall from a table and shatter into pieces but the pieces can never reassemble into the unbroken glass – you get the idea. This axiom seems to be consistent throughout the observable universe. One of the fundamental principles underpinning this is the *Second Law of Thermodynamics,* which states that the entropy of a *closed* system will (practically) never decrease into the future. Entropy is the measure of the *disorder* of a system. When returning to our egg (before we've made our omelette and assuming it has no cracks) we will observe an *organised system with low entropy.* When we make our omelette and crack the egg (two for me please) we create a *disorganised system with high entropy.* Left alone, entropy increases with the passage of time.

Just to complicate things a little, not all systems are closed and the second law doesn't deny entropy in open systems. Neither is it incompatible with evolution or complexity. To help clarify the differences between *open* and *closed* systems we can define each as follows:

Closed system - A physical system that does not interact with other systems. A closed system obeys the conservation laws in its physical description. It is also called an *isolated system.*

Open system - A physical system that interacts with other systems. The physical description of an open system is one that can appear to violate conservation laws; for example, energy will appear to be lost from the system of a motorcar engine over time, despite the law of conservation of energy. This is because the system is open, losing energy (in the form of heat) to surrounding systems (through friction). Earth is another example of an open system because it can exchange both energy and matter with its surroundings. A system that loses energy in this way also referred to a *dissipative system.* Understanding these systems is fundamental to comprehending the arrow of time and Professor Stephen Hawking (1942 -) one of the foremost contemporary theoretical physicists, expanded this idea by visualising time as a series of three arrows *(Fig 25).*

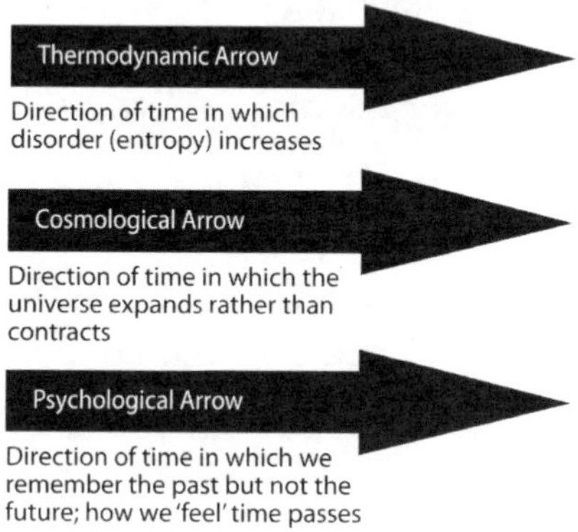

Fig 25. The 'three arrows of linear time'

The concept of these arrows helps to explain why we experience time as moving forward and how this 'flow' of time correlates with the expansion (and possible contraction) of the universe. Because the thermodynamic arrow of time is defined as the direction in which entropy increases, any universe with a thermodynamic arrow of time opposite to our own would become more orderly (less entropic) with the passage of time. However it is unlikely that life could exist in such a universe, because life as we experience it is based on the consumption of resources (more entropic).

The psychological arrow of time is the way that we perceive time - how we remember the past and anticipate the future. Science postulates that the psychological arrow of time must always point in the same direction as the thermodynamic arrow of time, because of the belief that consciousness is a direct result of chemical processes in our brain and bound by the *Three Laws of Thermodynamics.* If in some way this arrow were reversed with respect to the thermodynamic arrow, we would experience time flowing in the opposite direction - remembering the future and knowing nothing of the past.

The cosmological arrow of time points in the direction in which the universe is expanding. Some believe that unlike the thermodynamic arrow of time and the psychological arrow of time, this arrow could point in a different direction, particularly if the universe ceases to expand and begins to contract and collapse in on itself. Others believe that if this were to happen, the other arrows of time would change direction as well - entropy would decrease, and time would run in reverse, although we wouldn't know it since our minds would also run backwards. Hawking himself once believed that this would be the case, but has since retracted this idea.

All of this can be very confusing if consciousness and time are considered only as properties associated with the physical universe. If though, we take a few steps back and look from the level of infinite consciousness we will see that neither are bound by conventional scientific rationale.

Perspective once again plays a part in our understanding of time and as with many aspects of reality once we extend ourselves beyond our conceptual level of awareness, things appear so much differently. Quite clearly, there must still be something in place governing the functioning of life within higher realms but by the very nature of the ethereal existence this will encompass many contrasting qualities to those with which science is currently familiar. As we ascend higher up the planes of consciousness, time slows down while space becomes more flexible or amorphous and no longer possesses the fixed laws that we see here in the physical world.

Accessing global time from a linear perspective

Just because we experience time in a linear fashion here on earth, doesn't mean that we cannot also access information globally. Neither does our ability to do this contravene what we think we know about the flow of time or its direction. It does though, suggest that we are far more powerful than we yet realise in our natural ability to transcend the known limits of perception. When we access information globally, even though there is still some linear processing occurring, we are, through an expansion of consciousness transcending our usual limitations of receiving information. Returning to our DVD analogy, it is as if we are taking in the entire content of the disk in a snapshot along with the encoded meaning contained within it (we understand the movie completely) and then replaying it through our normal linear thought processes.

Another good analogy would be that of speed-reading. Anyone that can speed-read is able to scan the centre of a page from top to bottom and imbibe the entire contents in a matter of seconds. In terms of what consciousness can do however, even this seems pedestrian!

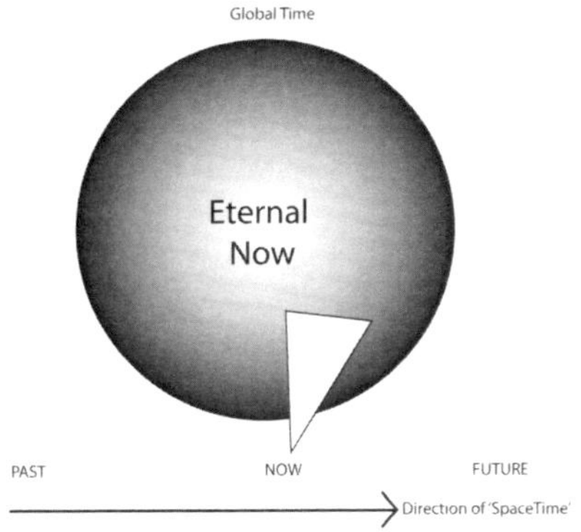

Fig 26. Accessing global time

We have all experienced instances when we experience greater awareness or insight into something - a kind of 'Eureka!' moment. This can happen at any time, not just when we are consciously attempting to expand our consciousness through processes such as meditation. It is as if we momentarily step into another dimension – and that is exactly what happens. The lens through which we experience reality re-focuses from linear time into global time and a whole new world becomes accessible.

Imagine being at the controls of the Hubble telescope and focusing in on a distant region of space or conversely looking through a powerful microscope into the realm of the infinitely small. This is what our consciousness does – sometimes involuntarily, but usually in response to a deeper intent or desire of the soul to connect with a greater field of knowing. When this happens, information pours into our minds, faster than the speed of light and we become instantly aware of it. Often what we access is the answer to something that we have been seeking or the knowledge that we have been searching for that takes us to a deeper level of understanding. It is through this process that some of the greatest ideas, discoveries and insights have originated.

Often, this ability to shift awareness and move into dimensions where time itself seems to expand occurs naturally during periods when a part of us temporarily leaves the body – such as when we are asleep (consciousness withdraws in part from using the body-mind computer) but it can also be made to happen at will. The seer, the remote viewer and the clairvoyant all share aspects of this ability to transcend linear space-time and *(Fig 26.)* shows this concept. The greater the spiritual evolution of an individual, the higher their energy curve and the more potent their ability to penetrate into the eternal now.

As a medium myself, I am often aware of being able to

access both past and future events, especially when connecting with someone for whom I am undertaking a reading. When this happens and a medium becomes aware of something that has yet to unfold within linear time, it has of course already occurred within global time and the sensitive is tapping into and decoding that information. We should not assume that all information obtained in this way is passed on from 'spirit guides' (although this can happen) but rather that it might be the expansion of the medium's own consciousness.

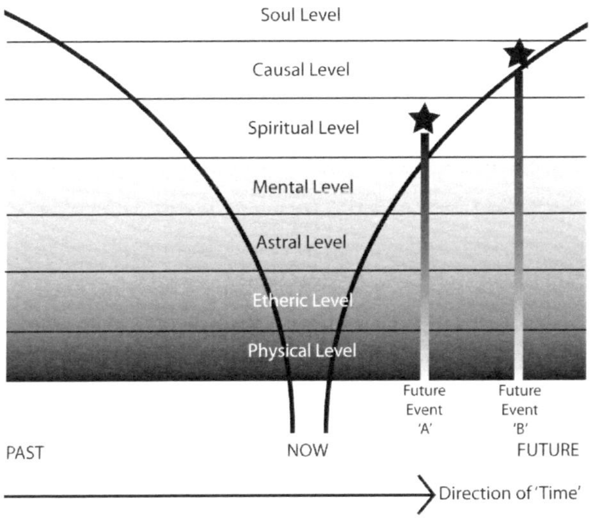

Fig 27. Accessing future events

In the above illustration we can clearly see that a consciousness possessing of the ability to either expand into the spiritual level or already residing there can see future event 'A' before it has occurred within linear time but may not yet be aware of event

'B' - also to occur a little further ahead within linear time. This is because they are unable yet to access the second event due to the fact that their consciousness and energy curve is insufficiently developed. Someone whose consciousness *is* developed to a greater degree or who resides on a higher plane of life will though be able to witness both events *before* they have unfolded in the 'now' of linear time and perhaps give a warning of what is to come. Such experiences, although considered by some to be supernatural, are the result of consciousness itself and are related to abilities that we all possess. Neither do they conflict with the forward direction of time or violate the laws of physics.

To illustrate how this works in a simple way, imagine riding a bike alongside a tall brick wall. You cannot see above the wall, but being the clever person that you are, you stand up (whilst still peddling) and peer over the top. From this perspective you can see what lies beyond the wall in several directions and can also see an intersecting road that crosses the path of your bike. A truck is travelling down the road (which you had been unable to see whilst in the seated position on the bike) and this information enables you to make a judgement of whether to continue peddling forward, or breaking to avoid a possible collision *(see Fig 28).* The astute reader will of course note that the cyclist, even though being able to peer over the wall and see what before was hidden, is still doing so within linear time and in the same way, when we are able to access global time we still have to process the information through a linear perspective, but the point is this; *through the expansion of our consciousness we are still able to access information within non-linear time* even though our brain decodes it in a linear fashion.

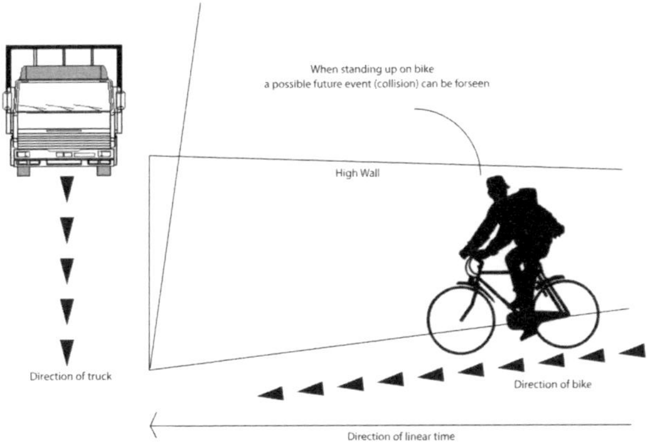

Fig 28. Accessing a future possibility

What matters most here is our ability to *connect* with information instantaneously when the blinkers are removed and our vision expands.

Time as a movement of consciousness

There is just no escaping the fact that time, space and consciousness are significantly inter-related and I will emphasise again that just as there can exist no time without space, so there can be no time or space without consciousness. The very fabric of space-time is intimately linked with consciousness at the macro and microcosmic scales and quite recently some theoretical physicists have begun to realise that not only does matter consist of energy, but also that energy consists of consciousness. When we let go of the grip of conditioned thinking we can begin to see how the apparently unshakable laws governing time and space are actually *relativities that respond to the movement of consciousness.*

The Indian philosopher Sri Aurobindo (born Aurobindo Ghosh, 1872 - 1950) writing eloquently in his book *The Life Divine (Book 2, Part 1, Chapter 2)* had a wonderful grasp of this concept:

"It would seem as if Time had no objective reality, but depends on whatever conditions may be established by action of consciousness in its relation to status and motion of being: Time would seem to be purely subjective. But, in fact, Space also would appear by the mutual relation of Mind-space and Matter-Space to be subjective; in other words, both are the original spiritual extensions, but it is rendered by mind in its purity into a subjective mind-field and by sense-mind into an objective field of sense-perception. Subjectivity and objectivity are only two sides of one consciousness, and the cardinal fact is that any given Time or space or any given Time-Space as a whole is a status of being in which there is a movement of the consciousness and force of the being, a movement that creates or manifests events and happenings; it is the relation of the consciousness that sees and the force that formulates the happenings, a relation inherent in the status, which determines the sense of Time and creates our awareness of Time-movement, Time-relation, Time-measure. In its fundamental truth the original status of Time behind all its variations is nothing else than the eternity of the Eternal, just as the fundamental truth of Space, the original sense of its reality, is the infinity of the Infinite."

Further to this, Aurobindo held the view that what he referred to as 'The Eternal Being' experiences three states of consciousness related to time. The first of these, referred to as 'Triple Time

Vision' and known in ancient scriptures as *Trikala-Jnana* is that of the timeless Eternal - existing outside or above the flow of time and not directly involved with it. The second, he described as a state of consciousness that can hold the awareness of the timeless state whilst also seeing the entire panorama of past, present and future simultaneously (Global time or the Eternal Now) *[my emphasis]*. The third state he described as the one with which we are all familiar – that of the flow of time from moment to moment, with the past receding and future events approaching. The present is like the blink of an eye and cannot be pinned down because it is always the future, becoming the past.

Bearing in mind what I have already outlined regarding our ability to transcend linear time, here is what Aurobindo had to say about the three standpoints that consciousness may take:

"For it can see the whole Time development from outside or from above the movement; it can take a stable position within the movement and see the before and the after in a fixed, determined or destined succession; or it can take instead a mobile position in the movement, itself move with it from moment to moment and see all that has happened receding back into the past and all that has to happen coming towards it from the future; or else it may concentrate on the moment it occupies and see nothing but what is in that moment and immediately around or behind it."

He went on to explain:

"This seeing of Time is not at all part of our normal awareness of events as they happen, though our view of the past, because it is already

known and can be regarded in the whole, may put on something of this character; but we know that this consciousness exists because it is possible in an exceptional state to enter into it and see things from the viewpoint of this simultaneity of Time-vision."

Another who understood the true nature of consciousness and its relationship to space-time was Mirra Alfassa, also known as 'the Mother'. Born in Paris in 1878 and a pupil at the Academie Julian, she became an accomplished artist, pianist and writer. As her interest in occultism grew she visited Algeria in 1905 and 1906 to study with the adept Max Theon and his wife. Her primary interest though, was spiritual development and in Paris she founded a group of spiritual seekers, giving regular talks to various groups. In 1914 the Mother voyaged to Pondicherry to meet Sri Aurobindo, whom she immediately recognised as the one who for many years had inwardly guided her own spiritual development. In April 1920 she rejoined Sri Aurobindo in Pondicherry and some years later when the Sri Aurobindo Ashram was formed he entrusted its full material and spiritual charge to her. When questioned about the nature of time and space, particularly relating to the higher frequency non-physical planes she had this to say:

"As there are forms, there is necessarily a Time, a Space, but it is not at all the same as the physical. It is neither the same Time nor the same Space. For example, as soon as you come to the vital world there is a Time and Space which are similar to the physical but without that fixity and hardness and irremediability which are here. That is, for instance, in the vital a

strong intelligent will has an immediate action; here, in the physical, it takes sometimes extremely long to be realised, an entire process has to be followed. In the vital it is direct, the will acts directly on the circumstances, and if it is truly of a very strong kind, it is instantaneous. But there is still a Space, that is, one has the impression of moving to go from one place to another, and that necessarily, as one moves, a certain time intervenes; but it is an extremely short time compared with physical time.

On the mental plane the notion of Time disappears almost totally. For example, you are in your mental consciousness, you think of someone or something or of a place, and immediately you are there. There is no need of any time between the thought and the realisation. It is only when the mind is mingled with the vital that the notion of time is introduced; and if they go down into the physical, before a mental conception can be realised a whole process is necessary. You do not have a direct mental action on matter. For instance, if you think of someone who lives in Calcutta, well, physically you have to take a plane and some hours must pass before you can be there; while mentally if you are here and think of someone in Calcutta, instantaneously you are there with him. Instantaneously, you see. But if you go out in the vital from your body and want to go somewhere, well, you have the feeling of moving, and of the time it takes you to reach the place you are going to. But it is incomparably fast in relation to the physical, to the time necessary to do things physically.

Only right at the top of the ladder, when one reaches what could be called the centre of the universe, the centre and origin of the universe, everything is instantaneous. The past, present and future are all contained in a total and simultaneous consciousness, that is, what has always been and what will be are as though united in a single instant, a single beat of the universe, and it is only there that one goes out of Time and Space."

When questioned on whether there exists a past, present and future upon the higher planes she answered:

> *"In the psychic plane? Yes, you have even the consciousness of all the lives you have lived. When you enter into contact with the psychic you become conscious of all the lives you have lived, it keeps the absolutely living memory of all the events in which the psychic took part – not the whole life, not that one can tell little stories to oneself: that first one was a monkey and then later something a little higher, and so on, the cave-man... no, no stories like that. But all the events of former lives in which the psychic participated are preserved, and when one enters into conscious contact with his psychic being this can be called up like a sort of cinema. But it has no continuity except in lives in which the psychic is absolutely conscious, active, permanently active, that is, constantly associated with the consciousness; so naturally, being constantly associated with the consciousness, it consciously remembers everything that has happened in the real life of the person, and the memories - when one follows these things – the memories of his psychic being are more and more coordinated and closer and closer to what could be a physical memory if there were one, in any case of all the intellectual and emotional elements of life, and of some physical events when it was possible for this being to manifest in the outer consciousness; then, at these moments, the whole set of physical circumstances in which one was is kept absolutely intact in the consciousness."*
>
> (Collected Works of the Mother, *Question and Answers,* 29 June 1955)

Quite clearly both Aurobindo and the Mother had a grasp of the concept of space-time-consciousness that I am advocating here including an understanding of how we experience the movement of time upon the higher planes of being. Imagine then, my delight at discovering this shared understanding, because at the onset of writing this book I was unaware of either of these enlightened individuals.

It's all happening now

If you are still having difficulty in grasping the concept of time (and I wouldn't blame you if you were) it might be helpful to take a sharp intake of breath, pause for a moment, step back and consider this; if the perception of time did not exist, at least at this physical level, what would be the purpose of existence? Because the movement of consciousness allows for perceptual space-time to be as it is and we experience our current reality as a series of occurrences happening *now,* yet appearing sequentially as a movement from apparent past to apparent present to apparent future, our life has order. Anything other than this, at least at this material level of being, would be chaotic.

Just as we exist within a wafer-thin band of visible light and sound and enormous areas of reality are tuned-out of our perception by our decoding mechanism for a very good reason, so is our limited awareness right here, right now, restricted for purposes that only become clear to us as we progress through the planes of life. The entire story of our experience of being human (we are much, much more than this) is one of restriction followed by incremental increases in learning and awareness. Were a small infant to begin life with all of the experience and wisdom of an adult, would this not remove the purpose and pleasure of childhood?

Consciousness grants each of us the delight of unfolding the inner beauty of our divine nature within an experience of time and all that it offers. We each have the joy and pain of memory, of forgetfulness, of reminiscence, of anticipation, of birth, of growth, of decay, of death and of everything in-between stretched out like sequins dotted along an invisible thread and all happening in this moment. To our mind it seems unfathomable and the more that we consider and attempt to reconcile it, the less that we seem to understand. Yet when all things are considered the wisdom of infinite consciousness has devised life in such a magnificent way as to allow our experience always to match our level of perception. If for now we are cognisant only of linear time and are unaware of anything more, then it is because this is the reality we are manifesting. When our consciousness changes, so does our reality and so does time.

For now, the majority of us remain within cyclic patterns, our lives interwoven by karmic events and soul choices that see us weave back and forth, alternating between physical on non-physical worlds and the rich and varied experiences that they offer. Throughout the course of our many lives, we spend a great deal of perceptual time expressing through one form or another as our journey unfolds and were it not for the existence of matter, we would not nor could not obtain all of the varied life experiences from which our ultimate spiritual wisdom is attained. We are diamonds waiting to be polished. We are golden ore, rough-hewn within stone and yet to see the light of day that brings our transformation into something beautiful and precious. We are the imprisoned splendour held fast by its own unknowing.

The physical universe – the universe of form, is what permits our very journey to take place and the memory of what we are and always have been, to be rekindled. But it will come as no surprise that even this celestial ocean may be far different than how humanity currently perceives it. What if we take a little look for ourselves?

Summary of main points:
- Time is not a constant
- Time is linked to perception and is a movement of consciousness
- Linear time is a facet of global time (the eternal now)
- Past, present and future coexist within this moment
- We can access global time from within linear time and witness events yet to occur at a particular level

Chapter Six

Constructing a Universe

"You are a function of what the whole universe is doing in the same way that a wave is a function of what the whole ocean is doing."

Alan Wilson Watts (1915 – 1973)

The universe – where do I begin? Such is the vastness of the ever-expanding physical cosmos that no words or descriptive text could really encompass totally its wonder or define its majesty. There are countless theories as to the origin and nature of the universe, each one as intriguing and captivating as the next but as we have already discovered, when viewed from alternate perspectives things appear differently. We have considered that on one level at least, the universe is holographic in nature whilst on another it is electrical. When viewed through the infrared spectrum its appearance is far different than when seen through a normal telescope and if an ultraviolet filter is applied, looks different again. Using red or green filters also modifies what we see as the information reaching our senses alters what we decode. When using telescopes astronomers try to look at the energy being produced by the universe not just in the visible part of the spectrum (what our eyes can see) but also through the entire range of electromagnetic radiation. Hence we have radio telescopes, infrared telescopes, X-ray telescopes and even some telescopes that are buried deep underground searching for particles such as neutrinos that are extremely difficult to detect. All of these processes deliver a view of the universe being a place of enormous mystery, which is without doubt true and seldom a week goes by without a new discovery of tremendous magnitude.

The human mind being what it is, desires more than it can ever quite attain and in relation to the nature of the universe this equates to an expanding number of theories about its conception, age, structure, shape, size, purpose, destiny and so on. Some of these are familiar to us, especially the aforementioned *Big Bang* theory that has been accepted by many as the most likely explanation of the origin of the universe. In this model our universe sprang into existence from a 'singularity' around 13.7 billion years ago. A singularity relates to a point where some property is infinite. For example, at the centre of a black hole, according to classical theory, the density is infinite (because a finite mass is compressed to a zero volume) - hence it is a singularity. After its initial appearance, the universe apparently inflated, expanded and cooled, going from extremely small and hot, to the size and temperature that it is now. This expansion and cooling process is ongoing.

The image *(Fig 29.)* published by the European Space Agency shows what is visible beyond the Earth to instruments that are sensitive to light at very long wavelengths. The bright horizontal line running across the middle of the image is our own Milky Way Galaxy's main disc. To suggest though, that this is an image of the 'entire universe' is rather like saying that the human genome offers a complete picture of humanity – neither is true.

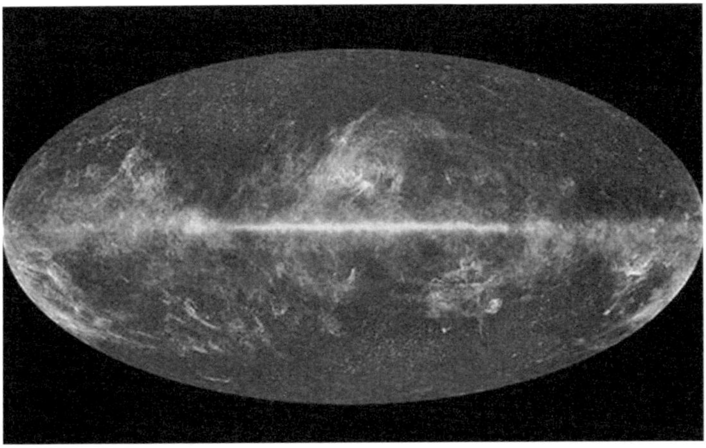

Fig 29. The 'entire universe' from Europe's Planck Telescope 2010 (photo ESA)

There are many misconceptions regarding Big Bang theory, the most common being that there was a gigantic explosion. Most experts agree however that this was not how the universe began but rather that there was and continues to be an expansion. One way of visualising this would be to consider an infinitesimally small balloon inflating to the size of our current universe.

Another misconception surrounds the tendency to imagine the singularity as a point somewhere in space, yet current thinking suggests that neither space nor time existed prior to the Big Bang. According to some calculations centered around Einstein's *Theory of General Relativity,* spacetime had a finite beginning that corresponded to the origin of matter and energy. The singularity may not have appeared in space but rather the other way around – *space appeared within the singularity.* What existed before this happened scientists can but speculate and the suggestion is that *nothing* existed.

It may not surprise you that I don't buy into this theory because there is little, if any evidence within nature of something having been created from nothing. The quantum physicist might disagree with this of course, especially when faced with the phenomena of sub-atomic particles blinking in and out of existence, but once again when we consider our multi-dimensional model of life we may see things differently.

From beyond the veil

The best selling author Lynne McTaggart (1951-) appears to share the same opinion when challenging the views of scientist Dr Richard Dawkins by stating: "This argument is simply scientific illiteracy. As any high school student of physics is taught, nothing comes from nothing." As she further points out, energy cannot ever be destroyed, it simply changes form, so the only universe that could ever emerge from nothing would be a universe without any energy which clearly is not the case regarding ours. Our universe is forever in motion, a swirling mass of subatomic particles that do indeed appear to pop in and out of existence but this is because of their interaction with a ground-state field - referred to as the Zero Point Field. The name originates from the fact that even at the absolute zero point, energy can still be measured. In her 2003 book *The Field,* McTaggart writes:

'Researchers discovered that the Zero Point Field contains the blueprint for our existence. Everything and everyone is connected with one another

through this field in which all information from all time is said to be stored. Ultimately, everything – from man to matter – can be traced back to a collection of electric charges that are continually in contact with this endless sea of energy. Our interaction with this field determines who we are, will become and have been. The field is the alpha and omega of our existence.'

Scientists have long understood that all elementary particles exchange energy with each other, combining and annihilating each other almost instantaneously causing random fluctuations of energy. These fleeting particles are referred to as 'virtual particles' because they only exist during this time of exchange and even in temperatures of absolute zero, the lowest possible energy state, where all matter has been removed leaving nothing left to create motion, these exchanges occur. This movement has caused some annoyance to scientists because it is ever present, rather like the background noise of traffic and messes with their equations. Because of this they often ignore or discount it and discount it - a process called 'renormalisation', enabling them to tidy things up and make everything fit – *except that it doesn't.* Just as *dark matter* and *junk DNA* are troublesome to cosmologists and microbiologists alike, subtracting the Zero Point Field is like removing God from the equation – it doesn't work. The theory of the all-encompassing Zero Point Field provides an unmistakable bridge between spirituality and science. Einstein couldn't prove but suspected it, when he said 'the field is the only reality'. This could explain the instantaneous 'ghost-like' transfer of information between quantum particles and also provide us with further insight into the origin of the universe.

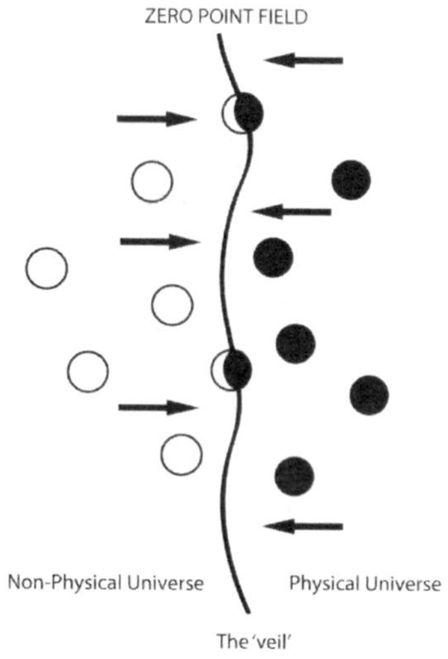

Fig.30 Virtual particles emerging from the Zero Point Field

As the illustration *(Fig.30)* shows, the permeable membrane-like 'veil' between non-physical and physical realities is not only a connective field but also acts as a kind of interface between dimensions allowing quantum interactions to occur and particles to switch 'sides' - appearing to blink in and out of existence. Central to this is once again our old friend *infinite consciousness.* For something, anything, to emerge there has to be something from which it comes and even virtual particles cannot be created from nothing. That 'something' is energy that exists beyond the physical frequency range and its momentum is a direct result of the manifestation of divine consciousness itself – the 'will' of God.

Now the big question is this; did the entire universe emerge in a similar way - through the Zero Point Field? I believe that it did. I also believe that it's emergence is ongoing and not a one-off event in the distant past. There are always clues that point us in the general direction of solutions if we recognise them. Nature is like this – providing answers to questions of enormous magnitude by giving us clues that are right before us. If we consider for example some of the patterns that occur throughout the natural world, often referred to as *sacred geometry*, we will clearly recognise both shapes and cycles that regularly occur within all levels of creation. The study of *fractals* or *fractal geometry* represents one such area of note and the famous 'Mandelbrot Set' *(Fig 31.)* is a clear example.

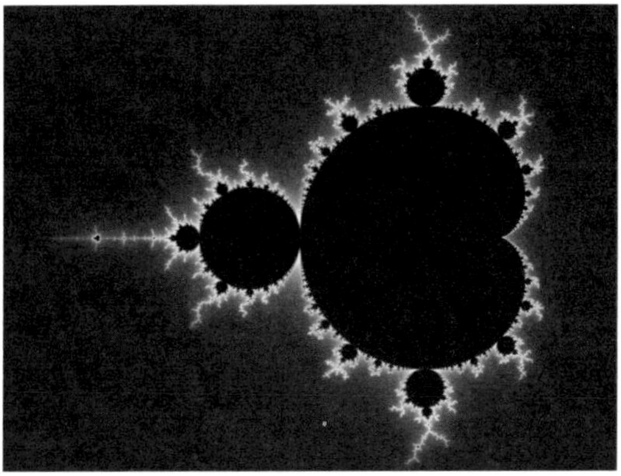

Fig.31 Mandelbrot Set

These geometrical blueprints structure all of creation. Each pattern forms an architectural blueprint for objects emerging within nature and each has its own unique resonance. They are also symbolic of the underlying metaphysical principle of the inseparable relationship that links the part to the whole.

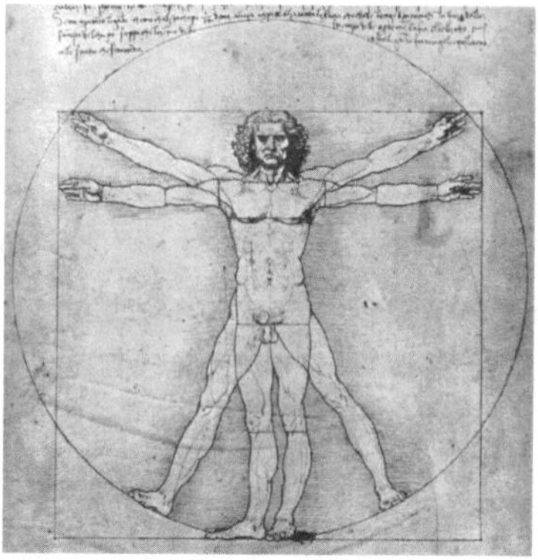

Fig.32 Leonardo Da Vinci's depiction of Man

It is well documented that many of the world's great works of art incorporate the principles of sacred geometry - such as Da Vinci's exquisite line drawing of the human form in *(Fig 32)*. His most famous work of art 'The Mona Lisa' is also said have been painted utilising sacred geometric proportions and many well known composers such as Mozart are also known to have structured their music by drawing on this knowledge.

The *Golden Mean* (also referred to as *The Golden Section, The Golden Ratio* and *The Divine Proportion*) is another representation of sacred geometry as is *Phi* and the *Fibbonaci Sequence.* This sequence of numbers, named after the Italian mathematician Leonardo of Pisa form part of series that runs as follows:

0, 1, 1, 2, 3, 5, 8, 13, 21 and so on.

Each number is the sum of the two numbers before it, after the first two numbers. Fibonacci first thought of the sequence as a solution to a problem he posed in a book he wrote in 1202, *Liber Abaci.* The problem was to discover the number of rabbits produced in a rabbit population beginning with a single newly-born pair - assuming that each pair produces another pair every month, they each become productive from the second month onwards and the rabbits never die. There are numerous connections between the Fibonacci numbers and the Golden Ratio. If we begin with a square of 1 x 1 then add another square of the same size and subsequently add squares whose sides are equal to the longest side of the existing rectangle we end up with a rectangle like this *(Fig 33).*

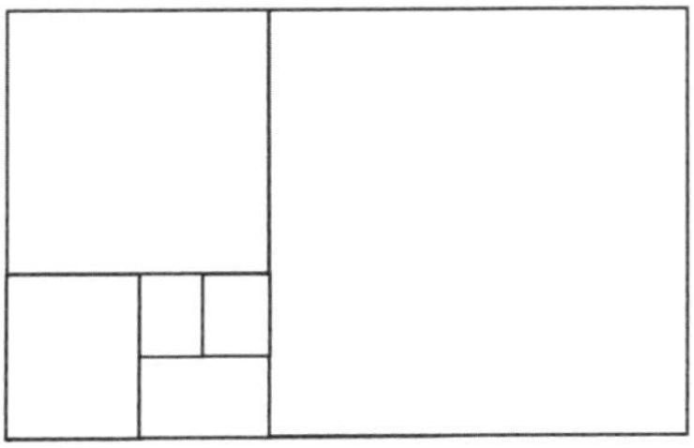

Fig.33 The Golden Rectangle

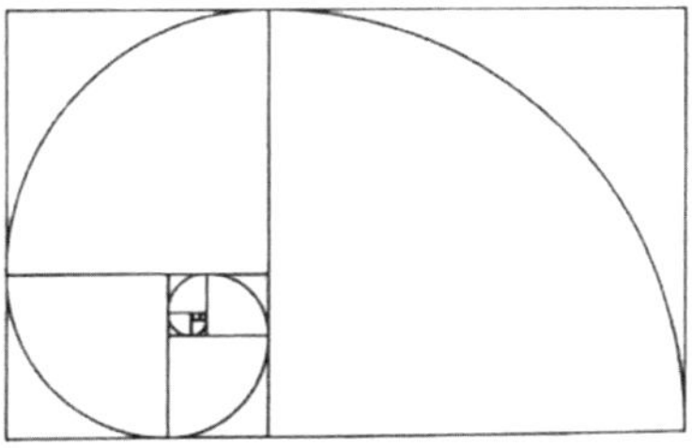

Fig.34 The Golden Spiral

If we now draw a quarter circle in each of the squares we get a Fibonacci Spiral or Golden Spiral as it is sometimes called *(Fig 34).* This is a pattern that you will find occurring frequently in many places throughout the natural world.

It is often said that 'God is a mathematician' and when we look at the many beautiful designs in nature we can see both the artistry and precision of the intelligence behind all form.

Here are a few obvious examples of spirals that correlate elements within our own world and the greater macrocosm beyond.

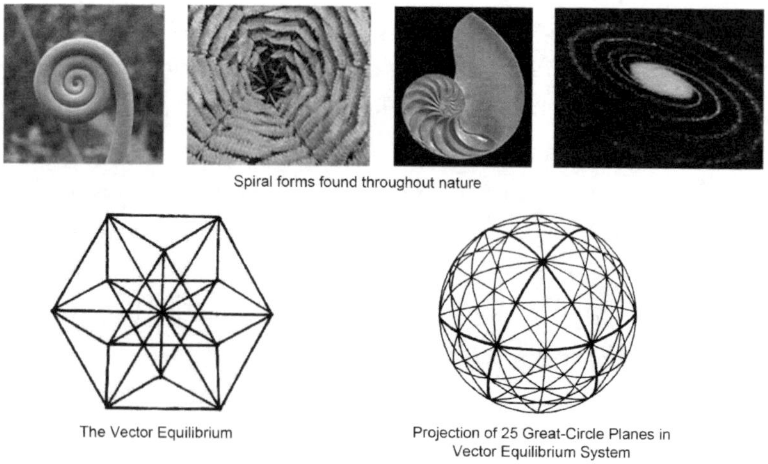

Fig.35 Nature's Golden Spirals (top) and the Vector Equilibrium (below)

One of the most significant patterns underpinning many forms within nature is called the *Vector Equilibrium (Fig 35)*. It is the blueprint by which nature forms energy into matter. A prolific inventor, Buckminster Fuller (1895–1983), coined the term because it is the only geometric form where all forces are equal and balanced. The energy lines (vectors) are of equal length and strength and they represent the energy of attraction and repulsion, rather like those felt around a magnet. In combination with the *Phi Spiral* and the *Double Phi Spiral* the 'VE' forms the basis of the most important geometric pattern found anywhere in nature – the *torus*.

The universe by design

In considering the origin of our universe it seems sensible to me to assume that the creative intelligence behind sacred geometry would also employ these same principles when designing the cosmos rather than leave things to develop randomly from the initial singularity. Was there indeed any one event at all, or simply the start of something that has continued ever since as witnessed by the expanding universe familiar to astronomers and cosmologists alike? If this is the case, what kind of sacred geometry could we expect the universe to have and why can't we see it? Seeing the 'shape' of the universe is extremely difficult because we are obviously inside it. This is a bit like asking a goldfish to describe the shape of its bowl. We can though, speculate and draw reasonable conclusions based around what we already know and understand.

Imagine for a moment that the beginning of the universe was not an explosion, but an ejection of extremely hot, condensed energy that surged forth from the singularity creating our version of space-time as it did so. Emerging through the veil between non-physical and physical realities this energy, travelling at enormous speed, but below the escape velocity of the system, spewed forth into the void trailing newly formed elementary particles in its wake. Gradually these individual particles began to cool, allowing clouds of hydrogen and helium to form and condense into stars which themselves either exploded or deteriorated expelling heavier elements into the cosmic dust of space, creating ever new stars that shone against the blackness of the interstellar ocean.

Eventually the jet of energy itself began to cool as it travelled from its source, slowing down due to gravitational forces, a process that continued until the cloud began to fall back and curl into itself like the head of a mushroom. Picture a fountain of water shooting upwards to a great height before falling back on itself and you will get a clearer image of how this would have appeared. Just as fountains often recycle their source of water so this flow of matter eventually found its way back to the source point, but not immediately. Such was its path that its own inertia took it beyond the source point only to be slowed down further by gravitational forces that drew it back again towards the centre.

In this ongoing process the numerous streams of energy collide, lose momentum and fall towards the nucleus in an ever-narrowing funnel of energy. This cycle is unimaginably long, encompassing enormous lengths of space-time and is also ongoing, continuing to this current moment. This is known as a *torus* and when seen from above it looks like this *(Fig 36)*.

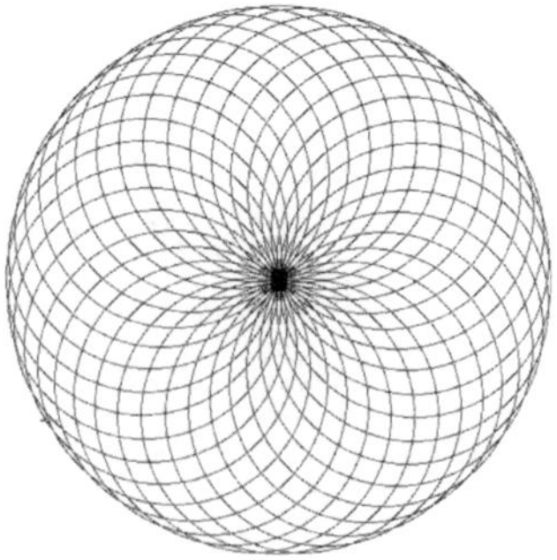

Fig.36 The Torus

A torus consists of a central axis with a vortex at both ends and a surrounding coherent field and has been compared by some to a smoke ring because of the curling, rolling motion that occurs. In both cases the energy circles back on itself to be reabsorbed into the cycle once more. There are again many examples of this energy pattern to be found in nature particularly in food and flowers *(Fig 37.)* whilst some mammals such as whales and dolphins are known to produce their own torus shaped bubbles in the water.

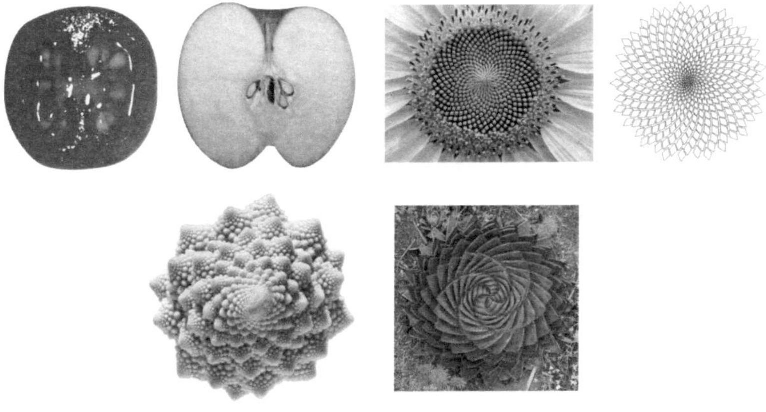

Fig.37 From left to right: Tomato. Apple, Sunflower, Sunflower torus illustration, Broccoli, Aloe plant.

The magnetic energy field around a bar magnet is also shaped like a torus as is the magnetic field of the earth *(Figs. 38 and 39 overleaf)*.

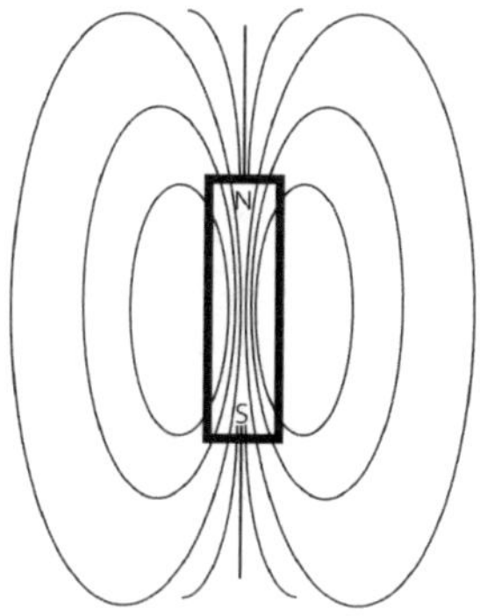

Fig.38 The energy field of a bar magnet.

Fig.39 The electromagnetic energy field of the Earth.

The concept of our universe as a torus may not be as inconceivable as some might suggest and one such thinker and visionary in particular - Itzhak Bentov (1923 – 1979), gives a stunning explanation of his theories in *Stalking The Wild Pendulum – on the Mechanics of Consciousness* a book that was years ahead of its time when published in 1977. Bentov, known for his exciting and original ideas arrived at the same conclusion of the universe being a torus and backed up his theory with detailed illustrations and commentary of how such a model could exist. In a follow up book *A Brief Tour Of Higher Consciousness* published in 2000 the author writes:

"Our universe has the shape of a torus, like an elongated hollow donut. In the centre of it is a focal point, a consciousness or the Creator represented by the white and black holes back to back. The Creator is like a seed containing all the information about the universe in potential form, as the acorn contains the pattern of the oak tree.

A jet of radiation emerges from the white hole and begins to flow around the Creator forming a skin or surface of the torus, which is the universe. Matter is born as it issues from the white hole, at first in the form of light radiation. As it cools down, stable particles are formed – protons, neutrons and electrons. The helium and nitrogen gases are formed. These condense into stars with heavier elements in their core. Stars collect into clusters or galaxies and as they die, they explode and create cosmic dust, which again condenses into new stars and planets. Everything in the entire universe, including our bodies is made of one cosmic substance."

Bentov, expanding on the theme of the torus universe explains his thinking behind the white hole/black hole theory:

"As the jet of matter continues to flow out of the white hole, it slows down and begins to be pulled back by the gravity of the original seed, the source in the centre. The jet reverses its direction and eventually is drawn back into the seed through the black hole. The gravitational collapse causes matter to be transformed into energy, as it passes from the black hole the

white hole, and it re-emerges from the white hole as a new universe. It is a continuous cycle of birth and death of matter.

One could call it a modest continuous bang universe, as compared with the Big Bang universe of a one-time explosion, spreading evenly in all directions."

Tying in this theory with the very nature of space-time and also revealing his thoughts about the holographic nature of the universe Bentov eloquently details his understanding of how the birth and death of matter in this model fit perfectly with quantum theory:

"In a continuous bang universe, the white and black holes in the centre of the torus are not only points of birth and death of matter, but also birth and death of time, for time appears where there are motion and matter. Time becomes a measure of the distance of matter from the point of its emergence from the white hole, and this distance reflects the stage of evolution of matter and its consciousness as it makes its cycle back to the source.

The universe is a hologram or interference pattern in which all parts are interconnected, containing information about each other and thus about the entire universe. As galaxies move along the surface of the torus, they radiate information waves about themselves. These waves interweave into an interference pattern and form a hologram, which fills the hollow space inside the torus, as well as extends into the surrounding void. This hologram of information could be called the Universal Mind, [Akashic Record - my emphasis] *because it contains all the information there is about the entire structure. Each individual human consciousness is part of that hologram, therefore, by projecting one's consciousness into the Universal Mind in a heightened state of awareness, one can obtain knowledge about the whole universe."*

Clearly, Bentov had a profound understanding of physics and combined this knowledge with his own spirituality, giving him a great insight into the nature and order of creation. He was amongst that most rare breed of men – the true esoteric scientist.

If we could view the universe from beyond the universe, the shape we would see, based on the torus model would be like this *(Fig 40)*.

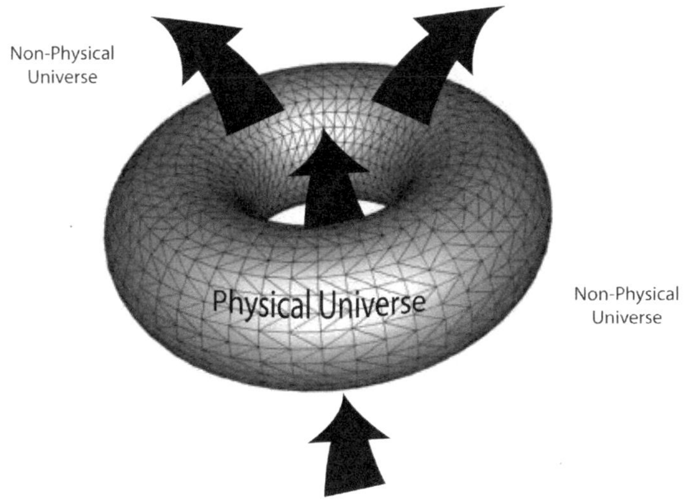

Fig.40 The universe as a torus

Now if you are thinking what I'm thinking when you see that shape, you are probably wondering what lies beyond it? In addition scientists often argue whether such a universe is a 'closed' or 'open' system. In some ways it is both, because all of the energy and matter produced eventually flows back into itself to be recycled again. Yet logic dictates that there must be something into which this ever shifting, ever evolving cosmos moves. Perhaps that something is the non-physical field of pure consciousness wherein exists all potential and out of which emerges all that exists including all of the higher frequencies and spiritual planes that we have detailed throughout this book – with the Zero Point Field representing the interface between the non-physical and physical domains.

In Bentov's model, there is also the suggestion that another kind of interface – the black hole/white hole vortex, exists. We could think of this as a type of cosmic plughole down which is sucked all physical matter and energy as it reaches the end of its journey, only to re-emerge somewhere else through the opposing white hole in what would be considered another Big Bang (or 'continuous bang' as he refers to it). This makes perfect sense to me and also typifies a fundamental principle within nature, that of cycles. Cycles are found everywhere we look, as are opposites. Is it inconceivable that the universe itself is part of a massive cycle encompassing birth – death – rebirth? It is beyond our comprehension to think that for every black hole with all of its apparent destructive tendencies an opposite white hole encompassing the essential creative forces of abundant life exists? I think not. Nature knows not the meaning of waste and as any scientist will testify, energy cannot be destroyed – it merely changes state. Bentov's sketch of how the entire cycle operated looked something like this:

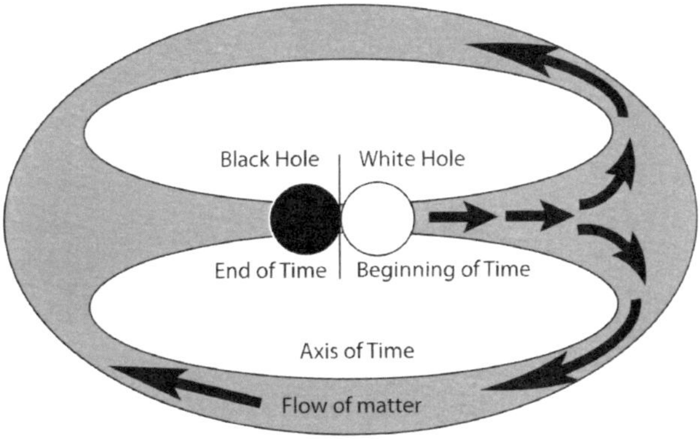

Fig.41 My interpretation of the torus cycle conceptualised by Itzhak Bentov

Being the inspirational thinker that he was, Bentov further postulated that our own observable universe (that which scientists can currently see) appeared like a bubble within the overall torus, much like an air bubble would look inside a glass of liquid *(Fig 42)*. This he suggested, would not only be expanding in all directions but also moving through the torus itself accompanied by a general expansion of human consciousness. As our universal bubble eventually approached the end of its life cycle and drew closer to the black hole, so there would be a subsequent decline also in human consciousness before the ultimate fate of falling into the gravitational singularity occurred.

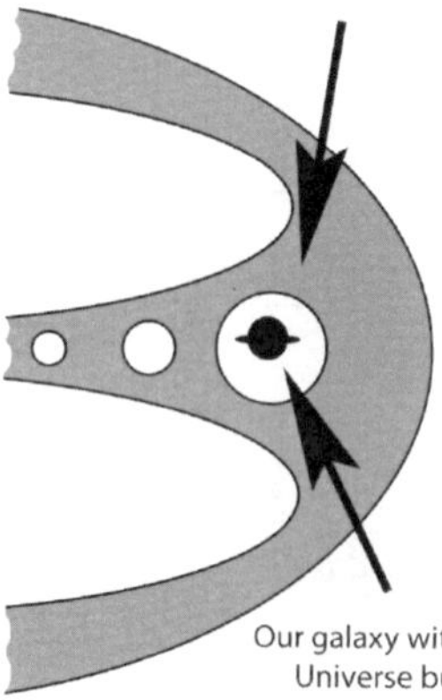

Fig.42 My interpretation of the Universe bubble within the torus conceptualised by Itzhak Bentov

It is my belief that this model of the universe is both viable and highly reflective of the intrinsic properties of nature itself, incorporating many if not all of the fundamental qualities found there. This being so, we can also recognise the same intelligence that resides within us, or that more accurately, *we ourselves are.* As each of us evolves to become self-aware, so the cosmos itself shares this same aspiration and it would not be unreasonable to suggest that entire galactic systems, encompassing billions of worlds are themselves intelligent 'systems' like ourselves. Some spiritual teachers have suggested that the earth for example, has a natural governing intelligence that is sometimes referred to as Gaia. The name originated from the shorter Greek word *Ge* from which the disciplines of *geo*graphy and *geo*logy were named and which refers to earth as 'the great mother of all.' In 1979 author

James Lovelock (1919 -) used the word Gaia for the title of his book *Gaia – A New Look at Life on Earth* which became an international best seller. With the advent of photographs of earth taken from space, humanity was able to see its home world as never before and in the book the author depicts Gaia as a nurturing entity regulating the climate and sustaining all life within her domain. Lovelock has since written further works based on the same principles outlined in the initial work but has commented almost apologetically, for his earlier somewhat romantic storytelling style when referring to Gaia in the female sense because of some unjust criticism from the scientific fraternity who suggested it was 'bad science'. Thus he has deliberately pursued an approach that employs the use of more scientific language and this can be seen in the second book *The Ages of Gaia.* The fact is though, that Lovelock hit upon something that captured the spirit of what many people have since come to recognise – *a sense of nurturing intelligence that is at work through every aspect of life.* This is what I have been detailing throughout this book and is *the same force at work throughout the universe,* at every level and every dimension. Surely if earth is an intelligent living entity then so must other worlds be? Infinite consciousness has to be unbounded consciousness and every facet of creation must exhibit the same qualities in degree – an essential principle of any holographic reality.

It is my suggestion that just as the patterns of nature repeat at both the microcosmic and macrocosmic levels, so too does the creative intelligence of all life reveal itself through the common thread that is witnessed wherever we look. If earth or

Gaia is an individual intelligence, whole unto itself yet part of the greater connective consciousness, then why not the other stars, planets and entire galaxies? Why not the universal torus itself?

Hidden in plain view

I don't know about you but if I were God, I'd want to leave some pointers to enable my creation to eventually work out what I was, where I came from and what sort of things I expected from them. Sure enough, that is exactly the way life is. The creator's fingerprints (and DNA if you like) are everywhere and there are some massive clues provided we recognise them. Like most things in life, when you are shown the way to do something or see something in a different light, the idea evolves into learning. Well, we are on a massive learning curve – one could even say a Fibonacci Spiral and we are honing in on the perfect ratio of understanding about the universe and ourselves. The blinkers are coming off and the remembrance of our true nature is being slowly revealed. Clues are all around us, some more obvious than others and provided we ask the right question, the solution will always emerge. The answer, inevitably involves consciousness. It has to, because everything *is* consciousness. So long as we factor this awareness into any model or equation that we devise – whether at the quantum or cosmic level, truth will emerge.

There will dawn a day when the very idea of the universe being considered as it now is by the majority of the scientific community will be as far removed as the common belief held in times past that the world was flat. The torus model is one of many, but I chose to include it because my instinct tells me that it represents something deeply significant. Perhaps the astrophysicists reading this will also consider it 'bad science' and question why I didn't delve into the realms of Superstrings, Brane worlds or other such theories? I remain open to all possibilities because I have learned that the more I know, the more I have yet to know but for now I choose to follow my gut instinct and my heart, for these even more than my head, speak their message loud and clear.

Summary of main points:

- Something cannot emerge from nothing
- The Big Bang may not have been a 'one off explosion' but an ongoing emergence of energy from the non-physical realm
- The Zero Point Field is the veil between non-physical and physical domains
- Nature reveals to us many structures and patterns known as sacred geometry
- It is likely that the shape of our universe conforms to one of these
- The torus model of the universe has distinct possibilities
- Space/Time and Matter emerge from 'White Holes' and return to 'Black Holes'
- No realistic theory or equation explaining the nature of existence can reasonably exclude consciousness

Chapter Seven

The Heart of Life

"There is an immense, painful longing for a broader, more flexible, fuller, more coherent, more comprehensive account of what we human beings are and what this life is for."

Saul Bellow (1915-2005)

The world that is our temporary home whilst we have chosen to undergo this experience of being human is often a chaotic place to live. Such is the gulf between ignorance and understanding that it is no wonder there are so many conflicts of opinion. Much of this human chaos is generated from the level of the lower ego-mind with a little help from the emotional level of course, because consciousness creates no such problems. Social structures, geographical location, history, poverty and wealth and a raft of individual and collective human traits centered around perceived power and control issues all play a part in ensuring that our short span here is difficult. Add into the mix nature itself and the normal rhythms and cycles of the earth and beyond, all of which we are a part, and the recipe for entropy is complete. Yet it doesn't have to be this way.

As we have demonstrated thus far, that which we undergo is of our own making. We are the architects of all that we experience as we pluck from the realm of pure potential our individual and collective reality and when *we* change, the world changes. Because everything is energy in motion and our human consciousness interacts with it - collapsing the waveform information into the holographic reality that we experience, we have a measure of choice as to what we 'download'. An evolved individual whose consciousness and energy curve enable a more refined interaction with life has the potential to create great beauty and wonder all around them whilst an ignorant mind will inevitably attract a lower energy frequency and attract a coarser type of reality.

The elements of fear, selfishness, egotism, intolerance, hatred and other negative states help to imprison human consciousness in a lower vibrational prison as well as generating incoherent energy waves that extend in all directions, creating further chaos. Imagine taking a stick and waving it violently to and fro in a pool of water and you would clearly see the surface become very choppy and agitated. Only when the cause is removed does the water return to its natural state of relative stillness and calm. This condition is *our* natural state too but because *disorder* is introduced through any number of ways the peace and tranquillity of our being is disturbed.

Those clever fellows the scientists have managed to record the heart rates of individuals displaying a range of emotions – everything from fear and anger through to happiness and love. The differences in the recordings is what one might expect – the more harmonious thoughts and emotions producing a regular coherent energy pattern as opposed to the more erratic path revealed by the disturbed energy of the agitated mind.

There is of course a wide range of mental and emotional

states, but as with both light and sound, the shorter frequencies are represented at the higher end of the spectrum. *(Fig 43.)* shows the shift in energy and coherence between the lower and higher emotional modalities.

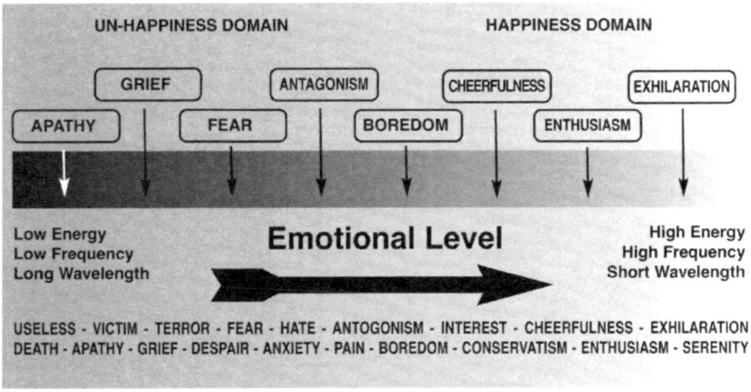

Fig.43 Mental and emotional states and their frequencies

What this represents is typified throughout nature. Higher states are always without exception found to incorporate finer, faster, purer forms of energy and it is no surprise that our thoughts and emotions operate in this way too. The implications of this, both for us as individuals and for the world around us are very powerful. Just as both words and deeds make an impact so do thought and emotion - both directly and indirectly. The old saying *'thought is the father to the deed'* is well known and often quoted and spiritual teachers down the ages have reminded us of the powerful connection between mind and body. What is now coming to light scientifically though is the way in which the electromagnetic energy fields of the mind operate in ways that have not previously been understood.

In Boulder Creek, California, an organisation known as *'The Institute of Heart Math'* has been carrying out some wonderful work in measuring the field generated by the heart and interestingly enough the *torus* emerges once more. *(Fig 44).*

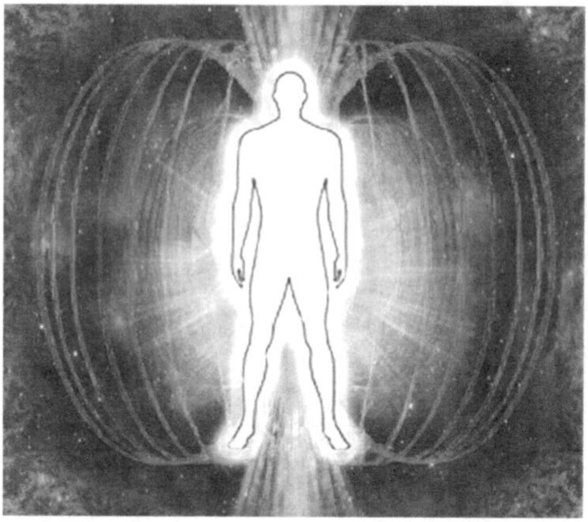

Fig.44. An interpretation of the torus shaped electromagnetic energy field generated by the heart.

The institute has undertaken extensive research in exploring the physiological mechanisms by which the heart communicates with the brain, influencing information processing, perceptions, emotions and health. It has long been recognised that intelligent mind exists throughout the body enabling consciousness to function upon many levels but new discoveries have revealed that the heart in particular is an organ in which mind functions in a profound way to influence both intelligence and awareness. In short, the heart seems to have a mind of its own. The heart, I would suggest, also has a *wisdom* of its own.

The intimate connection between the heart and the brain is striking in so many ways – in particular, those involving changes in mental clarity, creativity, emotional balance and personal effectiveness. The heart is much more than a simple pump circulating blood around the body – it is, to quote The Institute of Heart Math:

'a highly complex, self-organized information processing centre with its own functional "brain" that communicates with and influences the cranial brain via the nervous system, hormonal system and other pathways. These influences profoundly affect brain function and most of the body's major organs, and ultimately determine the quality of life.'

Quite powerful wouldn't you agree?

Heart intelligence

An early pioneer of research involving the heart and associated emotions, Walter Cannon (1871–1945) showed that changes in emotions are accompanied by changes in heart rate, blood pressure, respiration and digestion (that explains why when we are anxious, our heart pounds, we perspire, our stomach rumbles and we go red in the face). Cannon suggested that 'arousal' mobilises part of our *sympathetic nervous system* and prepares our 'flight or fight' response, whilst in quieter moments the *parasympathetic nervous system* has the opposite effect and calms us down. In his view, the *autonomic nervous system* and the associated physiological responses work in accord with the brain's response to a given stimulus and his assumption is that the brain is in overall control of the entire process. More recent research though, primarily by psychophysiological researchers John and Beatrice Lacey, challenges this view. The Laceys observed that this simple model only partially matched actual physiological behaviour. They discovered that the heart appeared to have its own kind of logic, distinct from the autonomic nervous system that seemed capable of sending purposeful messages to the brain that it both understood and obeyed. Even more significant was that it appeared that these could affect a person's behaviour. Further research has produced evidence suggesting that the heart and nervous system are not just acting on orders from the brain as Cannon had suggested, but are actively communicating with each other.

The emergence of *neurocardiology* has since provided many important discoveries and insights into the workings of the nervous system within the heart and how both the brain and the heart communicate with each other via the nervous system. The concept of the 'heart-brain' introduced in 1991 by another pioneer Dr J. Andrew Armour revealed a complex nervous system within the heart, sophisticated enough to qualify as a 'little brain' in its own right. The heart has also been classified as an endocrine or hormonal gland and is known to have the ability to produce and release hormones that exert effects upon the blood vessels, the kidneys, the adrenal glands and a number of regulatory regions of the brain. Other neurotransmitters once considered only to have been produced by neurons in the brain and ganglia outside of the heart have also been discovered to originate from a special kind cell type within the heart itself.

The ongoing work of The Institute of Heart Math and the latest research in neuroscience suggests that emotion and cognition can be considered as separate but interacting systems, each possessing its own intelligence. The heart appears to send a greater number of messages to the brain than vice-versa and it seems that emotions are even more powerful than thoughts, being able to displace mental thought activity. Thoughts on the other hand, do not appear to be able to so readily displace emotions. Emotions exert a powerful influence upon our cognitive processes and those most powerful of thoughts - the ones that are not easily dismissed, are often those in which emotion exerts the greatest influence.

Coherence between the brain and the heart or the mind and the emotions is key to the way in which we function upon this plane of life. When they are in-phase with each other our awareness is expanded and many of our cognitive abilities are influenced in a positive way. Conversely, when they are out of phase, these aspects are impaired. From this perspective if we consider ourselves to be a multi-dimensional information network (a biological computer) in which infinite consciousness operates through mental processes, emotional functions and physiological systems that work together we can see that for us to exist at the optimum level it is necessary for everything to work coherently. Thus, the heart, brain, nervous, hormonal and immune systems play vital roles in this process. When the 'conversation' between the heart and the brain is coherent the ability of the latter to decode information into the reality that we perceive is considerably enhanced. Worry, anxiety, nervousness and all forms of emotional stress change the signal from heart and affect the way that the brain functions. Is there anyone reading this who cannot bring to mind at least one personal experience when emotional upset clouded perception and the ability to 'think straight'? This quite clearly, is not just because of the brain-mind but also because of the heart-mind and the intimate relationship between both.

Within and without

Based upon what we know about human energy fields it is reasonable to suggest that the field radiated by the heart centre is able to exert influence not just internally but also externally. Because we do not end at the periphery of our physical form (the various electromagnetic fields in which our body is immersed extend for a metre or more beyond it - including the heart torus), we can safely assume that there are energy exchanges taking place consistently with people, animals, objects and places around us. If our energies are of a low vibration and our heart signal incoherent then we are in effect transmitting this into our immediate orbit, creating disharmonious conditions. Others, particularly those of a sensitive constitution, may well register this and be negatively affected by it. Consequently, their own bio-energetic system will be compromised resulting in the transmission of similar signals to others. Through this process a negative feedback loop is established and if persistent over time may even generate a negative resonant field, attracting similar thoughts and energies. If on the other hand, our heart is generating an ordered, harmonious field, this too being ever more powerful than an incoherent or negative one, will impact upon others who resonate with it. The implications of this are enormous because a truly loving, altruistic, selfless and spiritually centered individual will generate a coherent field that extends into the world, touching countless others in a positive way. They in turn, will sense this energy as it interacts with their own and a positive feedback loop is established *(Fig 45)*. The old paradigm of 'like attracts like' once more comes into play here as the power of the human heart works its magic.

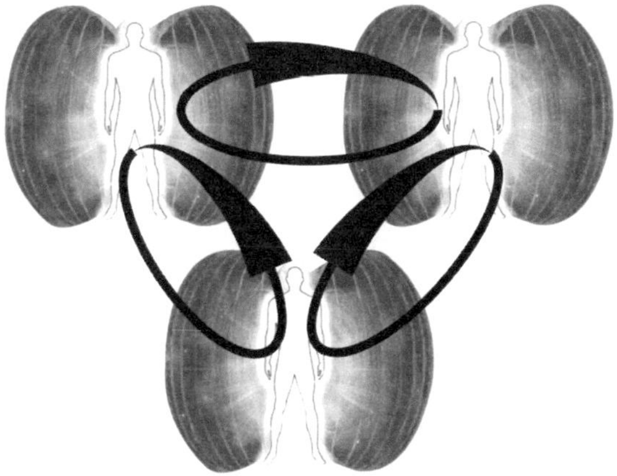

Fig.45. Coherent loops

The potential for the transformation of society through the power of the human heart is staggering. If only a small increase in the number of people generating a coherent heart field were to occur globally then positive changes would accrue. As things stand, there is so much fear, anxiety and apprehension in the world that the natural calm and peaceful state of human consciousness is being disturbed by a largely discordant signal being transmitted and recycled. A collective coherent energy has the capacity to reintroduce *order* into that chaotic system, restoring balance and wholeness.

From time to time we witness inexplicable happenings that epitomise the fragmenting of the normal human condition and that tell us something is wrong. Just as pain is nature's way of the body warning us that something needs to be addressed so extreme events, sometimes involving previously 'normal' individuals are symptomatic of problems lying deep within the human psyche. News stories of apparently sane people having committed shootings, mass murders or acting in bizarre ways are becoming more and more prevalent and it seems as if all rationality, morality and normality has deserted them. Are they victims of disturbances in their energetic fields that tip them over the edge? This is well within the bounds of possibility.

If we consider for a moment that we are at one level at least, electrical in nature, existing in an electrical universe and a world in which we are bathed by both natural and man-made electrical energy fields, then is it not possible that our human circuitry could become overloaded or even re-wired in some way? Of course it is. If you input too high a voltage into an appliance designed to operate at a lower power level, you either blow the fuse or burn out the circuitry and prevent the unit from working. This is why we have fuses and circuit breakers that act as safety nets when the system is overloaded – and humans operate in a similar fashion. When someone loses it, or 'blows a fuse' this is exactly what has occurred.

Geopathic Stress

A brief word here about *Geopathic Stress*. The term is not a new one. 'Geo' means 'Earth' (or Land) and the term 'Pathic' indicates both a disease, as well as a cure for disease. The most accurate definition is the study of earth energies and their effect on human wellbeing. Compounding the problem, we have surrounded ourselves with an ever-expanding plethora of devices and installations that are increasing year upon year. As *(Fig 46.)* reveals these largely man-made energies surround us almost to the point of being unavoidable and in combination result in immersing us in an 'electro-magnetic smog'. It seems as though common sense regarding the potential harm that can be caused by surrounding ourselves with so many electrical fields has been largely ignored and a wider understanding of human energy systems is needed. Other external environmental stressors can also impact on human health and push individuals beyond physiological tolerance and a growing body of scientific evidence is revealing many links between mental and emotional attitudes, physiological health and long-term wellbeing. A broader recognition and understanding of the heart-mind to brain-mind connections can only help facilitate positive change.

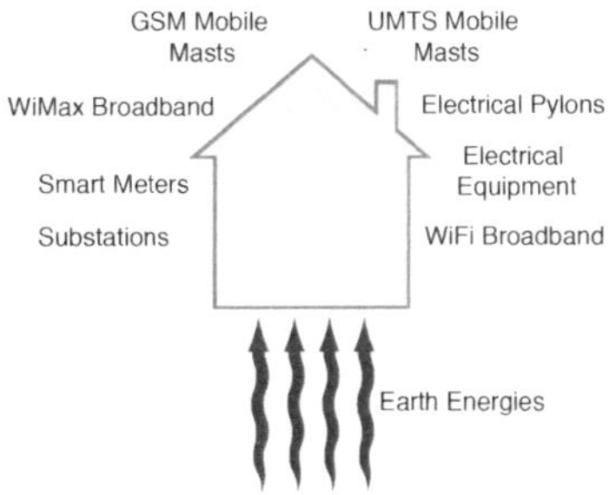

Fig.46. Geopathic Stress hazards

Geopathic stress zones also impact on us in ways that are little understood by the majority of people although the science of *Bio-geometry* is helping to address this. Just as we have our own human circuit board known as the *Meridian System* so the earth's energy grid can be thought of as a web that holds or links the world together. As discussed in Chapter 3, the system of ley lines and other geometric shapes that intersect the globe are also responsible for the movement and release of energy. These include *Curry lines,* and *Hartmann lines* both of which supposedly play an important role in the natural functioning of the planet.

Curry lines are said to be a global grid network of electrically charged lines of natural origin that run diagonally to the poles and were first discovered in 1951 by Dr. Manfred Curry (1899-1953) - founder of the Bioclimatic Institute in Bavaria and a Dr. Wittmann. There is some disagreement as to how wide apart these lines are, but the general consensus seems to be around 3 metres - although most experts recognise that this can vary. The lines themselves are not seen as problematic, only the points at which they cross. Naturally lines that run in this fashion will have numerous intersecting points but because the lines are electrically charged, the intersecting points are either double positives, double negatives or one of each. From his studies Dr. Curry speculated that the positively charged spots lead to a proliferation of cells, with the possibility of cancerous cell growth, whereas the negatively charged spots could lead to inflammation. Similarly the Hartmann lines or *The Hartmann Net* consists of naturally occurring charged lines, running North-South and East-West named after Dr. Ernst Hartmann (1915-1992), a German medical doctor. Again we find alternate lines that are usually positively and negatively charged and that are said to criss-cross the globe. Where they intersect it is possible to have double positive charges and double negative charges, or one positive and one negative charge and it is the intersections that are seen to be a source of potential problems.

My own view on earth energy grids is that as a general rule, nature does not evolve permanent fixtures that harm life. Of course, there are instances of highly destructive events such as earthquakes, volcanoes, tsunamis, pole shifts and the like, but these are usually 'one off' occurrences. For the very structure of the planet to encompass an energy grid that is, under normal circumstances detrimental to its cohabitants, seems highly unlikely. However, it is quite possible and indeed likely that human intervention and meddling could have impacted upon negatively over time to interfere with the normal operation of such a system. As with our own circuitry, blockages can and do occur resulting in a variety of problems.

Balancing the breath

Whilst control of the breath during meditation states and as an aid to facilitate spiritual development is well known, the practice of *coherent breathing* is a relatively new development. It has been recognised for centuries that breathing is the key to health and wellbeing but once again the term 'coherence' emerges and the latest studies reveal the power of certain breath techniques. By breathing 'coherently', we can consciously modify the operation of our autonomic nervous system and the result is that each time we inhale, our entire being tends to flex whilst each time we exhale our entire being tends to relax. The net effect is *balance.*

There are many techniques employed throughout various spiritual practices and belief systems and almost without exception they lead to more coherent heart rhythms and thus coherent energy fields in humans. Whatever label we choose or method we select, the end result is more harmony and balance. Above all else, this is what is needed most in the world today and we have the power to create this. It is sometimes said that *'The mind is king of the senses but the breath is king of the mind'* and there is a great deal of sense in this. We *'breathe to live, but do not live to breathe'.* However, when our breath is regular and rhythmic and we become conscious of the energy that we inhale and exhale we are accessing a great power. Here is one very simple technique that you may find helpful. I refer to it as *'Coherent Cosmic Breathing'.*

Coherent 'Cosmic Breathing' technique

1) Find a comfortable place to relax, where you will not be disturbed.
2) Switch off your phones (including any mobiles).
3) Breathe in and hold your breath for two seconds. As you do so feel that you are holding life-energy in your heart centre.
4) Continue normal breathing but do so, slowly and quietly. If you can leave a small gap between inhalation and exhalation, do so, but only if you feel comfortable with it.
5) As you continue to breathe slowly and gently, allow your thoughts to focus of the purity of the breath and know that you are breathing peace into your body. As you exhale, breathe out

the opposite of peace which may be considered as restlessness. Again, as you breathe in the positive force of peace and you feel the gentle love that accompanies it, so you breathe out the negative force of restlessness and notice any fear or anxiety that leaves your body with it.
6) Now, feel that you are breathing in cosmic energy with each breath and that this gentle universal force is purifying your mind and body, flowing like a river into every part of you. As you breathe out, any impurity is leaving and your whole body and mind is in harmony.
7) When you have practiced this many times, notice that your breath and the cosmic energy is coming in and going out of every part of your body – through your heart (especially) through your eyes, your nose, your ears and the pores of your skin.
8) Feel the joy in your heart and the energy field around it, shaped like a torus, expanding way beyond the periphery of your body, reaching out into life and projecting a coherent energy into the world.

It is known that some spiritual masters are able to breathe even with their nose and mouth closed and when you have practiced this technique and mastered it, much of your impurity will be replaced by the light and power of infinite consciousness. Rhythmic patterns of slow, deep breathing strengthen the respiratory system and help to calm the nervous system. When the breath is controlled, the mind naturally follows and when the mind is subjugated through the control of the breath, the body follows by releasing anxiety, stress and tension. Problematic thinking then ceases, cravings and desires diminish and we are no longer prisoners, giving us the freedom to embark on our path towards spiritual liberation.

Coherent meditation

Many people across the world meditate and the process of mediation, like that of conscious breathing, takes many forms. Controlled breathing is synonymous with meditation but not necessarily essential to it. One can meditate very successfully without changing the breath or even being consciously aware of it. Whatever form mediation takes - and it can be anything from simple contemplation to entering a deeply altered state of consciousness, there is a movement of energy and a positive change in both the heart field and the electromagnetic force within the auric fields. Meditation is a healing mechanism for both the individual and for the world and I can think of no genuine form of meditative state that does not of itself produce a coherent field.

Prayer is also a form of meditation and true prayer (not that which is repetitive and stereotyped) is an extremely powerful force for good. How often do we hear it said 'Pray from your heart'? When we pray selflessly for others, perhaps when we are genuinely concerned and really need divine intervention to bring about positive change for a situation that seems beyond help, we are sending forth coherent energy into the resonant field that connects us. In addition, we are also linking with the non-physical realms to connect with the healing energies of life itself and those that channel them through us to the source of their greatest need.

Some groups suggest their members' link together at specific times of the day to help connect them to the source of healing and help. One such organisation is *'The Harry Edwards Healing Sanctuary'* based at Shere in the UK. They have their 'healing minute' twice daily at ten in the morning and ten in the evening when people from countries around the world join with the sanctuary to focus their thoughts on healing for those in need and for world peace. Perhaps the most recognised and well observed accounts of large numbers of people coming together in prayer, again held in the UK is when on November 11th each year at 11am the country falls silent as people remember in prayer those who have passed during conflict. This is known as 'Remembrance Day' or 'Armistice Day'.

Similar unifications occur regularly when at sporting events and suchlike, people are remembered during a 'moment's silence' or sometimes a 'minute's applause'. During this time everyone, almost without exception unites as one mind or a single connective energy in contemplation and consideration of the said individual or collective. Surely there is more to this than simply the power of suggestion or command? When the announcer says over the speaker system that the minute silence is about to begin, there is seldom any question that everyone will obey – it just happens. Yet there is something more profound occurring here that deserves a fuller explanation and that is connected to *group coherence.*

Group coherence

Earlier we mentioned resonant and morphic fields and

the ways in which they connect individuals. Something similar occurs when large numbers of people gather together and share a common perspective. Close physical proximity is not required, just a similar focus of consciousness at a given time. That time can be prearranged (as in the healing minute) or designated there and then (a moment's silence before the start of the football game). Either way, the directive is given and people follow by linking in to the flow of energy created. Some will do it because they feel it is expected of them, others because they genuinely feel a desire to do so but whatever the reason, a consensus is formed and an unspoken agreement undertaken. Some might suggest that this is the power or suggestibility of the mind, but I think that it is much more than this.

I am advocating here that during a purposeful coming together of this kind there is a shift in consciousness and for a short time at least, *an opening of the consciousness to a higher state.* I have myself been in such situations; particularly at large football stadia amongst perhaps a hundred thousand spectators, at least half of who 'oppose' the team that I 'support'. Yet during the designated silence, rivalry ceases, we become 'one' and proud to be so. In that short space there is a unifying presence and the realisation that we share something deeper together. In these moments I have also sensed a certain pride at being able to share with everyone in the stadium thoughts and emotions of a corresponding nature – revealing the extraordinary power of empathy and love towards our fellow man. At times like these a great coherent field is created and were it possible to record this it would I suggest, appear momentarily as a giant torus field extending across and beyond the arena.

I cannot emphasise enough the power that is generated when people unite through a positive and purposeful intent. The potency of a single individual whose consciousness and mind are working benevolently towards a positive outcome for others is a delight to behold, but when the collective energy of many hearts and souls combine, an irresistible force is unleashed. Mountains can be moved, barriers torn down, obstacles removed, illusions shattered and paradigms shifted. The power I'm highlighting here isn't physical of course, although at one level it can be. Rather, I'm referring to the flow of goodness and positivity, that when harnessed flows from the hearts and minds of unified individuals into the collective heart and mind of humankind. This is coherence at its most transformational and unstoppable.

The Earth's 'heartbeat'

I could not complete this chapter without mention of another deep connection that we each have with our planet and it is here again that the nurturing spirit of Gaia weaves its way back into our thoughts. We are quite simply, intimately connected with each other, with the earth and with the greater universe in which we exist. We reflect the nature of our reality and our reality is a reflection of us. When we resonate together in harmony, life for the most part flows and we flow with it. When there are disturbances of any kind – mental, emotional or physical, these are felt in degree at both a global and individual level. The famous line in the Star Wars films *'may the force be with you'* is very apt here because we can often sense subtle changes in the force of the planet and of life itself before any noticeable physical changes take place. Animals for instance, are known to be able to sense an impending earthquake and birds have an in-built sense that enables them to navigate across enormous distances with great accuracy. Before the devastating 2004 Indian Ocean earthquake and subsequent tsunami that claimed the lives of over 150,000 people in eleven countries it is claimed that many creatures either fled or moved to higher ground and there are numerous accounts throughout recorded history of such occurrences.

We too are connected in numerous ways to earth energies, some of which I have already outlined in this book. One of the most powerful of these connections is with the Schumann Resonance, named after the German Professor, Winfried Otto Schumann (1888 -1974) of the Technical University of Munich who successfully predicted their presence in 1952. Schumann asserted that there existed electromagnetic standing waves in the atmosphere within the cavity formed between the earth and the ionosphere *(Fig 47)*.

Fig 47. The Schumann Wave

The discovery came about when during a lesson about ball condensers Schumann asked his students to calculate the frequency between the inner and outer ball, equating to the earth and its ionosphere. The figure they arrived at was 10Hz and two years later in 1954 this was confirmed when further experiments detected resonances at a main frequency of 7.83Hz.

There is a spherical 'cavity' between the earth and the ionosphere created by the conductive properties of the earth's surface and the outer boundary of the ionosphere, separated by non-conducting air. Electrical discharges such as those produced by lightning spread into the cavity producing electromagnetic waves in the very low frequency (VLF) and extremely low frequency (ELF) ranges. Although there are fluctuations in the resonance – radiation from the sun being one such cause, the frequency of 7.83Hz has been verified by many researchers as being accurate. When Schumann's test results were published, a physician named Dr Ankermueller immediately made the connection between Schumann Resonance and the frequency of brain waves. He contacted Schumann who was intrigued and asked a doctorate candidate, Herbert König, to investigate further. König began a series of studies to compare human EEG recordings with natural electromagnetic fields in the environment and was soon able to confirm the relationship between the earth's resonance and human brainwaves with the frequency of 7.83Hz being a key factor.

Another foremost researcher, Dr Wolfgang Ludwig came up with an idea to take more accurate calibrations and devised a method of measuring from both above and below the earth's surface. During his research Ludwig came across the ancient Chinese wisdom of the Yin (feminine) and Yang (masculine) energies and made the connection with the strong signal of the Schumann wave surrounding the earth (Yang) and the weaker geometric waves coming from below the surface (Yin). The teachings state that for health and harmony to exist the two signals must be balanced and Ludwig discovered that this is indeed the case.

In 1963, Professor Rutger Wever, from the Max Planck Institute for Behavioural Physiology in Erling-Andechs, constructed an underground bunker that screened out the earth's resonance. Wever was engaged in experiments with light and frequency and studying the affect on circadian rhythms (Circadian rhythms are physical, mental and behavioural changes that follow an approximate twenty-four hour cycle, responding fundamentally to light and darkness in an organism's habitat). In each experiment, student volunteers lived there for four weeks in a hermetically sealed environment. In the subsequent study results, Wever noticed that when Schumann Resonances were filtered out of the bunker, the students' physical and mental health suffered. They all experienced varying degrees of stress and migraine headaches but incredibly, after brief exposure to the 7.83Hz frequency that had been screened out, their health rapidly improved. The first astronauts who, having moved beyond the range of Schumann waves were no longer exposed to their influence encountered similar problems and modern spacecraft are designed to counteract this issue with the introduction a magnetic pulse generator that mimics the earth's frequency.

Perhaps *Geopathic Stress* is also having a bearing on our connection with the Schumann Resonance, especially with the advent of wireless technology and the electromagnetic soup in which we find ourselves. These man-made signals are virtually drowning out the natural signals that have existed for aeons of time and which may act as a kind of 'tuning fork' for our own natural biological systems as well as those of all life on earth. The danger, particularly with microwave signals such as those used by mobile phone technology and which are pulsed at frequencies close to the Schumann Resonance is that we may be creating a hostile environment that is literally `out of tune' with nature itself.

There exists a need for more research into the connectivity between all life on earth and the planet itself if we are to continue to thrive in an ever- changing environment and questions must be asked regarding what dangers are posed by man-made pulsed frequencies and how they affect the Schumann Resonance, which at the time of writing has reportedly increased in some areas to 12Hz. Some researchers suggest that this is giving the effect of time 'speeding up' - although there does not appear to be any scientific evidence to corroborate this, whilst others have also linked this increase to a shift in the earth's magnetic poles or even a 'flip' - either of which would have implications for life on our planet. If living organisms, including humans, do respond to and perhaps even depend upon electromagnetic fields and earth energies – even weaker ones, then it makes good sense to investigate further. We simply cannot ignore the fact that at this level at least, we are 'electromagnetic beings' sensitive to electromagnetic radiation and dependent upon existing within energy fields that suit our entire constitution. Suitable frequencies can improve and enhance life whilst harmful ones can impair or destroy it. Quite simply, we have to maintain the coherence between the planet, our environment and ourselves if we are to continue to function normally. The choice is ours.

Summary of main points:
• We exist within an ocean of electromagnetic energies
• Emotions and thoughts generate energy fields within and around our body
• The human heart is capable of generating a very powerful coherent field
• The human heart has its own intelligence
• The coherent field produced by the heart connects us with others
• Through these connections, 'coherent loops' are formed
• Man made 'electronic smog' impacts upon our wellbeing
• Coherent breathing can help to restore harmony
• Coherent meditation enhances our connection with others and helps raise our consciousness
• The earth has it's own 'heartbeat' with which we are intimately connected

Bruce Tainio of Tanio Technology in Cheney, Washington developed equipment to measure the biofrequency of humans and foods. He used this to determine the relationship between frequency and disease. See Appendix I (Fig 54) for a table charting the correlation between electrical frequencies and health/disease.

Chapter Eight

Emerging Horizons

"We have always held to the hope, the belief, the conviction that there is a better life, a better world beyond the horizon."

Franklin D Roosevelt (1882-1945)

Relationships form a fundamental part of our existence, naturally occurring throughout creation, encompassing everything from the parasitic to the symbiotic, from the selfish to the altruistic and all that lies in between. Nothing ever exists in isolation. Everyone and everything is connected at the core level of being and the idea of separation, imposed by the collective world-view and fuelled by the ego mind is but a persistent illusion. Infinite consciousness, focused through the lens of the human mind within the realm of matter, knows this, yet such is the powerful nature of the illusion that it is unable for the most part to express anything other than a fragment of its totality.

There is some evidence to suggest that there was a time in ancient history when the level of human consciousness was greater than it is now and that following some kind of massive upheaval it began to take a downturn. Scholars suggest there was indeed a 'Golden Age' when a kinder, gentler more benevolent human race existed and war and conflict were largely absent and emotions like hate and greed unknown. Legends that speak of places such as Atlantis and Lemuria describe a world of great abundance and beauty whose citizens lived lives of a far higher order than we do.

Clearly something occurred to change the order of things with the likely candidate being a catastrophe of global proportions that resulted in movements in the landmass of the planet. Although the earth has undergone many upheavals over the course of its life, geological and biological records suggest that as recently as between 11,000 and 13,000 years ago a huge cataclysm occurred. Whether this was the result of a meteor strike or some other naturally occurring event such as a pole shift we can but speculate, but what seems certain is that a single happening or series of events of enormous consequence affected the course of human evolution.

Our enquiring human nature leads us to question why such a significant occurrence took place and also what life would be like now had it not. How would we be acting towards each other today had our pathway not been fundamentally altered? How would our society be structured? What aims and aspirations would we have? How spiritually advanced might we be? What inventions and discoveries would we have made? What cures for illness and disease would we have found? What inherent abilities would we have unfolded?

Yet we need not overly concerned, because if like me, you intuitively sense a greater intelligence at work, you will have no doubt realised from all that I have outlined to this point that infinite consciousness will not be thwarted or denied its highest expression whatever setbacks it encounters. Even if we humans have wandered off track or regressed in any way, we can be sure that our inner compass is still pointing us in the general direction of progress.

In his book *On The Origin Of Species,* Charles Darwin (1809-1882) outlined his theory of evolution, suggesting that populations evolve over the course of generations through natural selection. This idea, similar to the way in which science has embraced the Big Bang theory of the origin of the universe, seems to have been generally accepted as accurate. It is my contention though, that there also exists an intrinsic *spiritual* purpose in operation and that the selective process is not merely the result of adaptations occurring at the physical level or through 'the survival of the fittest' mechanism but also encompasses an unseen supernal component operating from the highest level of being. Indeed, nothing in my experience suggests that events are ever simply the result of the interaction of material forces or physical laws but are inevitably part of a greater scheme of which we are seldom aware. The intelligence out of which all emerges and is ever a part operates with a ceaseless perfection which neither natural nor artificial intervention can thwart. Being bound by neither time nor space and operating across all frequencies and dimensions the *will* of the creator is manifested. The words found within the Lord's Prayer 'Thy will be done' take on a deeper meaning when viewed from this perspective.

We each share a relationship with the unmanifest as it works through and within us because *we are that.* Our *individual consciousness* is part of *infinite consciousness* and as a consequence we cannot fail, either individually or as a race despite any setbacks we encounter along the way.

Paradigm shift

The term 'paradigm shift' is an over used one, yet it can be realistically applied to what is already underway on this planet. In the opening chapters I outlined how consciousness, operating through the mechanism of the brain, decodes waveform information into an illusory (but very real from our current perspective) holographic reality. Well quite simply, when our consciousness changes or expands, then the waveform information that we access and the reality we experience also changes. Human cognitive ability is shifting and expanding. It is transforming itself to engage deeper levels of awareness and understanding. *Transcognitive Spirituality* embraces the shift in consciousness that incorporates the integration of the scientific and esoteric paradigms into what can be called *the science of spirituality*. It offers a more comprehensive description of the true nature of reality than conventional approaches and in addition offers insight into the mechanisms of paranormal phenomena so often ignored by mainstream science *(Fig 48)*.

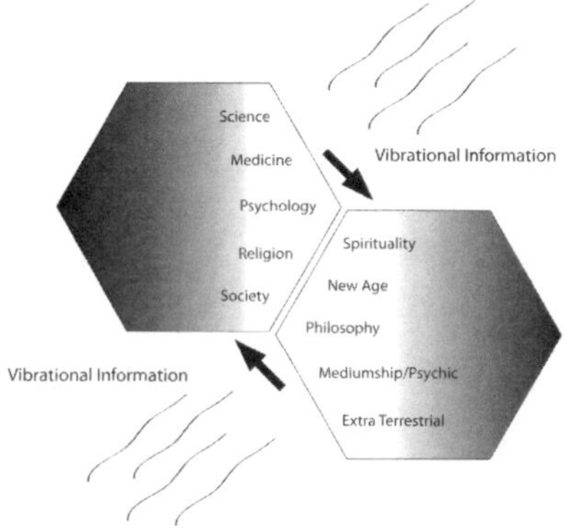

Fig 48. Transcognitive Spirituality – the unification of science and spirituality

With the advent of global communication and particularly the Internet, groups and individuals worldwide can more easily and quickly access information, allowing an exchange of ideas and beliefs and the greater integration of disciplines that once seemed poles apart. This, coupled with our deeper connectedness at the level of the collective unconscious means that more and more of us are able to develop an awareness and gain a grasp of concepts that would once have been beyond the understanding of the layman. Who for example, would have been able to comprehend anything at all about the behaviour of sub-atomic particles, say fifty years ago, unless they had themselves worked within that field? Today, even some school children have a rudimentary understanding of quantum physics and this is happening in many areas of life with the onset incremental learning from an ever-earlier age.

As any sentient species develops, changes occur on all levels of being – physically, mentally, emotionally and spiritually and what is termed 'progress' can be seen to occur through linear time. At any point there are extremes within a group, ranging from the underdeveloped to the very developed. Usually these two extremes exist as minorities, with the majority lying somewhere in between. As time passes and the race evolves, so what was the minority group (underdeveloped) progresses to become the majority (norm), whilst the current majority also progress to where the very developed group once were (new norm). In the meantime the former minority of very developed individuals have also progressed themselves to become highly advanced souls *(Fig 49 overleaf)*.

As the illustration shows, at any given period upon the physical plane there will always be groups of individuals who are less evolved and more evolved than the majority, with the under-developed group always playing catch-up and the majority group similarly striving to reach the level of the more highly evolved. Thus, the gulf in awareness existing between opposite ends of the developmental spectrum is vast, which is why there exists so much ignorance, pain and misunderstanding. The good news though, is that with any paradigm shift, awakening occurs and even though there is inevitably some discomfort as the old gives birth to the new, a greater understanding eventually emerges.

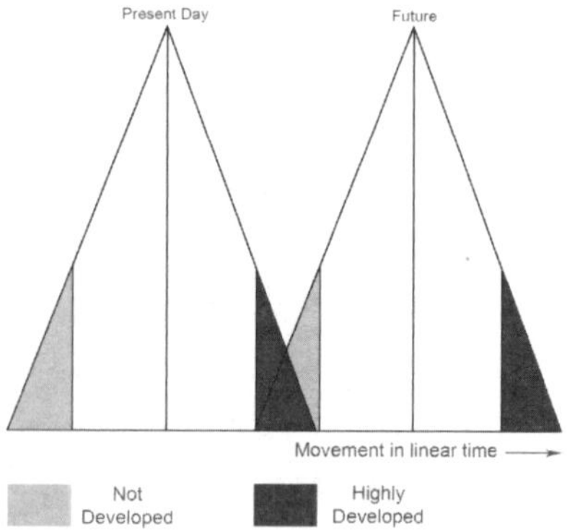

Fig 49. Progression of the race

It is fair to assume, based on the understanding that infinite consciousness is wholly benevolent, that the natural tendency of every highly developed species is to reflect abundantly the qualities of its creator. Indeed, I cannot conceive of any truly evolved race anywhere in the universe that could behave any differently – otherwise they could not be considered as such. As any wise soul will tell you, truth is always greater than ignorance and light more powerful than darkness and these innate principles reside within the essence of life. Just as language was an inevitable consequence of the very structure of DNA, so truth and wisdom are hard-wired into the fabric of evolution. It can be no other way – infinite consciousness says so.

So where does this leave mankind, right here, right now?

Precursors of change

Evolution can be defined as *the change in the inherited traits of a population of organisms through successive generations.* When living organisms reproduce, they pass on to their descendants a collection of traits, some of which may be obviously discernable, such as the colours in a peacock's tail or the number of spots on a leopard, but they also include characteristics based upon the sequence of nucleotide bases that make up their DNA. In fact, when we talk about evolutionary inheritance, it is DNA that we are actually referring to - the transfer of genetic sequences from one generation to the next and when particular genetic sequences change in a species (for example through mutation) and these are inherited across successive generations, this is the true stuff of evolution. Some researchers tell us that there is no 'end goal' of evolution but that it is a series of inherited traits occurring over time resulting, almost inevitably in more complex biological forms. It is also important to clarify the distinction between natural selection and evolution because they are not entirely the same thing. For example; trait changes among the members of a population are not always a result of selective processes but may be attributable to other influences. However, my own view is that there is a higher directing influence at work, a kind of driving force, directing in a purposeful way the upward movement of humanity. I call this the *higher divine principle* and it operates from and through the higher frequencies of being, eventually filtering down and through to the physical level within the DNA.

There are always harbingers of change amongst us and it is not unusual to find mutations of form occurring within a small number of individuals before extending to the majority. Just as a virus may first invade one person and then spread to others, so genetic change, driven by the spiritual directive is seen first to manifest through a small select group. Such individuals are inevitably regarded as odd or abnormal and may be ridiculed or ostracised by their peer group and society at large. In ancient times, some noted philosophers, scientists and mediums amongst others were punished for their beliefs and ideas that were seen as evil, wrong or simply against the norm. In recent times there have been striking examples of individuals, often young children or teenagers who display abilities and share tendencies not seen in the majority of people. Sometimes they are described as 'gifted' or 'special' as in the case of so called 'Indigo children'.

Savants

Savant syndrome is a condition in which a person who is considered severely challenged in their ability to communicate exhibits exceptional ability or brilliance in a particular field. Often such gifted individuals are on the autistic spectrum and some have either brain injuries or neuro-developmental disorders. The term '*idiot savant*' (French for 'learned idiot' or 'knowledgeable idiot') was first used to describe the condition in 1887 by a British doctor, John Langdon Down (1828-1896), better known for his description of a more common genetic disorder known as Down Syndrome. The use of this term lasted for a while but has gladly now been discarded. More recently another phrase *'autistic savant'* has been employed but like 'idiot savant', is now viewed as something of a misnomer because only one-half of those diagnosed with savant syndrome were found to have autism. Upon realisation of the need for greater accuracy of diagnosis and dignity towards individuals, the more acceptable term '*savant syndrome*' is now widely used.

Often, savants display mental ability that can only be described as 'super-human', demonstrating incredible feats that are seemingly beyond the capability of the majority of us. Imagine being able to play a complex piece of music after hearing it only once, doing difficult mathematical equations easily in your head or being able to remember and reproduce precisely in a drawing every window in every building having flown over a city the size of London. This is what a savant is capable of and what they lack in social interaction or normal cognitive functioning almost pales into insignificance when compared to this kind of ability.

Questions inevitably arise; does the 'normal' human brain have latent savant-like abilities? Do our higher cognitive functions somehow block this capacity, and if so how and why? Is it possible that we too could be endowed with remarkable savant-like abilities without the accompanying autism or developmental disabilities? And most significantly, is it *we* (the majority) who are actually the ones impaired? That last question may not be as ridiculous as it first appears.

In one study by Allan W. Snyder (1940 -) of 'The Centre for the Mind' in Sydney, Australia an attempt was made to simulate the rare form of brain impairment found in savants in healthy volunteers by directing low frequency magnetic pulses into their left fronto-temporal lobe, effectively impairing its operation. The results, although not conclusive (others have had mixed results in trying to replicate them) indicate that savant-like abilities may indeed be artificially induced by simultaneously stimulating the right side of the brain and inhibiting the left using low-frequency electrical currents. This is not totally surprising as the right side of the brain is the source of human imagination and creativity and many artists, thinkers, philosophers and mediums are predominantly right-brained. The right hemisphere of the brain is also able to look at the 'whole' (global time, the eternal now) whereas the left hemisphere looks at parts and is more analytical (linear time and the present moment).

There are many extraordinary savants, some who do not have notable developmental problems and a few whose abilities are so exceptional (these are known as *'prodigious savants'*) that under any circumstances they can be classified as phenomenal. Here are just three amazing individuals deserving of a mention:

Stephen Wiltshire (1974 -) – the 'human camera'.
As an infant, Stephen was a mute, having been diagnosed as autistic. He attended a school for special needs children and there he discovered a passion for drawing. First he drew animals, then London buses, then buildings and the city's landmarks. Throughout his childhood he communicated through his drawings and

slowly with the help of his teachers, he learned to speak and by the age of nine uttered his first word 'paper'. Stephen has a particularly striking talent: he can draw an accurate and detailed landscape of a city after seeing it just once. He has been featured on several TV programmes and is famous for drawing an accurate 10 metre long panorama of Tokyo following a short helicopter ride over the city complete with every skyscraper and building. In a 2001 BBC documentary entitled 'Fragments of Genius', he was taken on a helicopter ride over London and three hours later, he completed an accurate sketch of four square miles nearly perfect in perspective. Now an adult, Stephen has four books to his name and has opened a gallery in London where he presents his artwork each week.

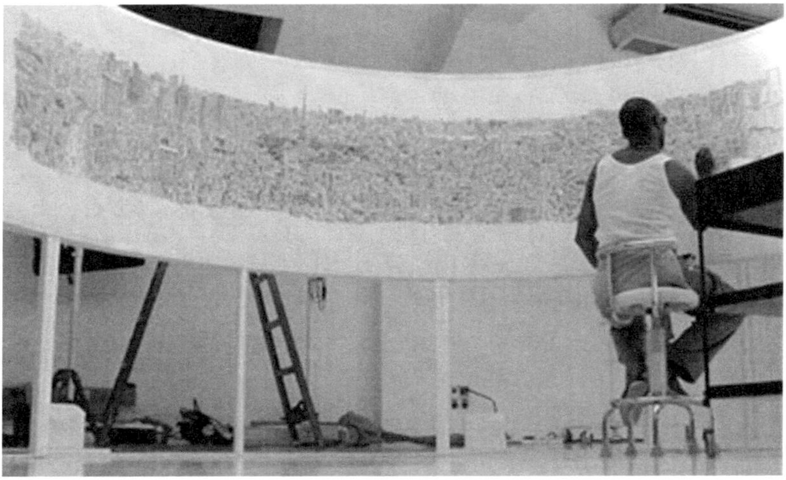

Fig 50. Savant Stephen Wiltshire at work.

Here are some of his truly amazing drawings of cities across the world.

Fig 51. Top (left to right) Birmingham, England and Sydney, Australia Bottom (left to right) Manhattan, USA and Rome, Italy

Leslie Lemke (1952 -)

Leslie Lemke was born with severe birth defects that required doctors to remove both of his eyes. His own mother surrendered him for adoption, and a kindly 52 year old nurse named May Lemke who already had five children of her own adopted him when he was just six months old. As a young child he could not swallow and had to be force-fed until he learned how. He was also unable to stand until he reached the age of twelve. However, when he was just fifteen something remarkable occurred. One night, May awoke to hear the sound of the piano playing and upon investigating found Leslie flawlessly playing Tchaikovsky's Piano Concerto No.1. He had no previous musical training and had only heard the piece once on a TV programme. Following this, he began to play a variety of pieces and styles and only had to hear something once in order to be able to perform it. Like Stephen Wiltshire, Leslie appeared on television and as his fame spread, played many concerts around the world.

Fig 52. Savant Leslie Lemke – 'a true musical genius'

Orlando Serrell (1968 -)

In 1979 a normal Orlando Serrell, then aged ten was playing baseball when the ball struck him hard on the left side of his head causing him to fall to the ground. He eventually got up and continued playing but soon began experiencing headaches. After these had subsided he realised that he could perform complex calendar calculations and could recall the weather, where he was and what he was doing on every day since his accident. Following growing media interest he appeared in various local newspapers and then later on NBC News. His brain functions were also investigated through an MRI scan and used to compare against non-savant individuals. He has also appeared on the Discovery Channel.

Fig 53. 'Acquired Savant' Orlando Serrell – 'calendar brain'

One explanation of such extraordinary ability might be the result of a past-life proficiency that has somehow transcended the normal timeline sequence, crossing over into the current incarnation and that savants are actually 'old souls' displaying gifts from a former life. This may indeed be true as at one level all information is holographically stored and under certain circumstances can be retrieved, but perhaps there are other reasons. One possibility is that such individuals are the forerunners of change within the human race, acting like advanced messengers at the leading edge of evolutionary development – although I think this unlikely due to the associated developmental disorders that so often accompany savant syndrome. A second, more likely reason and the one that I believe is true is that savants exhibit abilities that lie dormant within us all. What people like Stephen Wiltshire, Leslie Lemke, Orlando Serrell and many others reveal are glimpses of the natural ability of humanity that has become atrophied across time. The gifts that they demonstrate reveal to us the awesome power of *infinite consciousness* operating through mind and *this is our natural state of being.* It is we who have become impaired and for whatever reason our natural functioning denied. This is without doubt a powerful statement to make and the logical, left brained majority will question and perhaps deride the very notion of the idea. Yet as we have already seen, nature has ways of communicating to us if we are open to them.

When we consider what is actually taking place within the brain of savants we are given an insight into the difference between our true nature *(individualised consciousness part of infinite consciousness)* and the vehicles that it employs - namely the mind operating throughout the brain and the physical body *(bodymind).* Because the physical brain of a savant is damaged or impaired in some way, the cognitive processes of the mind are forced to operate differently. Savant skills however, far from being unique are I believe, possessed by everyone.

"It's not that savants are cleverer than the rest of us," says Snyder, "it's just that most of us go one step further in our brain processing—from detailed facts to meaningful concepts - and once we've done that we can't go back."

In the brain of a 'normal' person every tiny detail of say, a visual image is first registered, then processed and finally edited, with most of the fine detail removed, all within the space of less than one second so that what then appears within our consciousness is presented as a single useful idea (see chapter one for more information on how the brain works). In short, the stuff we don't consciously need to know is edited out, leaving just what we 'need' to know. This is a little like the news headlines and sound bites that we are exposed to through the various media outlets. There are reasons why this occurs of course, mainly as a result of human evolution and the survival instinct – if a large and ferocious tiger is running towards you there is no need to count the number of stripes on its back, you just get the hell out of its way! In the case of savants this suppression doesn't happen, so the picture that they see appears in fantastic detail, with every component in place.

Using the same reasoning, Snyder believes that if you or I were asked to discern the precise day of the week that corresponded to a particular date – for example what day the tenth of March 2040 falls on, we would have great difficulty because knowing that information would have little practical use to us and would therefore be edited out before passing into consciousness. A savant whose particular obsession was within this area though would not encounter this problem, being able to arrive at the answer easily and quickly.

Another clue as to whether savant-like abilities are inherent within us all may lie within the natural ability of young children. Imaging studies of brain activity in newly born infants shows that it is limited to regions that we are unconscious of as adults but which register sensory information that provoke automatic responses, urges and emotions. The area of the brain associated with conscious thought and perception known as the cerebral cortex becomes more active as a child grows and increasingly handles the processing of information. Around the age of eighteen months when the rudimentary signs of speech begin to emerge this process is accelerated and this may help to initiate activity in the frontal cortex where conceptual processing mainly occurs. In autistic children however, this shift appears to be either slowed down or incomplete allowing their savant-like qualities to be retained.

Another researcher, Wisconsin psychiatrist Darold Treffert (1933-) asserts that autistic people often show both structural and functional dysfunction in the left hemisphere of the brain. Once again we see that the right hemisphere is allowed greater dominance resulting in the qualities of this aspect of mind and consciousness to be displayed. In his books *Extraordinary People: Understanding Savant Syndrome*, and more recently *Islands of Genius* Treffert also explores the phenomena of genetic memory - instances in which individuals somehow "know" things they never learned and also "acquired savantism"- where a neuro-typical person unexpectedly and spectacularly develops savant-like abilities following a head injury or stroke. He writes of a reservoir of untapped potential or 'an inner savant capacity' lying within us all. Of course this potential lies within us all, because I state again - *it is what we are at the core level of being.*

Sensitive souls

Another development to occur in recent times has been the emergence of so called 'Indigo' children. This is a term that I have to say I find irritating and one that initially smacks of new-age quackery along with similar titles such as 'Crystal' children, 'Rainbow' children, 'Star' children and 'Blue Rays'. It would though, be all too easy and unfair to dismiss these fanciful titles as being simply alternative labels coined by those not willing to acknowledge a range of conditions affecting children such as Attention Deficit Hyperactive Disorder (ADHD), Attention Deficit Disorder (ADD), Obsessive Compulsive Disorder (OCD), Dsylexia and other Learning Difficulties.

Clearly, something is happening here that is impacting on children worldwide. I have often thought that changes in DNA are being reflected in the way children are today, particularly in regard to behavioural problems and there is some evidence to suggest that this may be attributable at least in part, to food additives and other man-made pollutants. Yet I cannot help but ponder that something deeper and more profound is also occurring that again, is linked to the paradigm shift that is underway.

Suppose we selectively ignore the new-age terminology and connotations to various high profile disorders for a moment and consider replacing these with the term *highly sensitive person?* This immediately reframes our view and puts a different slant on things. What if highly sensitive people were themselves the forerunners of evolutionary change – part of nature's response to the way in which humankind has been behaving over the last few centuries toward itself, the planet and all creatures upon it?

One person who has already focused on this is author Elaine Aron who has written a series of books including *The Highly Sensitive Person* and *The Highly Sensitive Child.* Her self-help works explain the phenomenon of SPS (Sensory Processing Sensitivity) and how this affects around 15% - 20% of the population. She suggests that SPS is no illness or syndrome, but is an 'inborn temperament' of the HSC (Highly Sensitive Child). As Aron explains on her website:

> *"This makes them quick to grasp subtle changes, prefer to reflect deeply before acting, and generally behave conscientiously. They are also easily overwhelmed by high levels of stimulation, sudden changes, and the emotional distress of others. Because children are a blend of a number of temperament traits, some HSCs are fairly difficult - active, emotionally intense, demanding, and persistent - while others are calm, turned inward, and almost too easy to raise except when they are expected to join a group of children they do not know. But outspoken and fussy or reserved and obedient, all HSCs are sensitive to their emotional and physical environment."*

Does this all sound familiar? Do you have a child that encompasses these traits or perhaps know one that does? The chances are that you do and as the author also points out:

> *"Unfortunately, the trait has been somewhat misunderstood in our culture, so that most psychologists and parents tend to see only one aspect of some sensitive children and call this trait shyness, inhibitedness, fearfulness, fussiness, or "hyper" sensitivity. If one could see inside the mind of a sensitive child, however, one would learn the whole story of what is going on - creativity, intuition, surprising wisdom, empathy for others..."*

In addition to these, the aforementioned 'Indigo children' also tend to exhibit other traits such as sleep disorders, insomnia and persistent nightmares. They are generally very energetic individuals and may be extremely sensitive towards environmental toxins, developing food allergies amongst others. They may also appear introverted and react strongly to loud noises, anger and negativity.

Whilst it is important to keep a sense of perspective in regard to 'special' people or those who appear different from the norm, we should never be too quick to dismiss other possibilities and to take a step back in order to view the larger picture. In my own work as a trance medium, an area in which I have been involved for over forty years, I have learned to respect the need to look at life from more than one viewpoint. As my dear wife will also testify having listened to an 'other-worldly' intelligence speak through me on many occasions, things can appear very differently when witnessed from another perspective.

Mediumship

The gift of communication between the ethereal and physical realms has played a significant part in connecting our higher and lower selves across time. This natural ability often referred to as mediumship is without doubt one of the most significant ways in which we connect to the non-physical frequencies, operating in varying degrees of accuracy according to individual ability. The world of the medium though, whilst often appearing glamorous to outsiders, is anything but. As with the sensitive child and the savant, a natural medium seldom enjoys an easy life pathway. Sensitivity is a double-edged sword and true mediums seem to suffer as much as anyone, often beset with earthly problems, many of which are health related due to the constitution of their human make up and the vehicle that allows their gift to function. Over use of the nervous system particularly through 'physical mediumship' where ectoplasmic energies are employed can have a detrimental effect on an individual and problems involving the endocrine system in particular may develop over time. I also know from personal experience that for the sensitivity of any medium to be awakened, some prior form of mental, emotional or physical 'suffering' inevitably occurs. It is as if pain itself is the catalyst that opens up compassion, sensitivity and the desire to serve the needs of others through the highest power known and attainable to mankind. Indeed, before anyone criticises the ability of any medium, they would do well to consider the path that has led them to develop their gift and the true purpose behind their work.

Mediums are, to use a well-worn phrase 'on a hiding to nothing' when it comes to proving life after death. Connecting with planes beyond our own is never easy and is not, cannot and never will be an exact science. Each attempt to bridge the gap in frequency between earth and the next level is an experiment that can never be exactly replicated. The old saying 'a medium is only as good as their last sitting is so true. I have witnessed many fine demonstrations of mediumship over the years, but equally have seen and heard some absolute dross from those whose reputation is highly respectable within the spiritualist movement. It isn't that they have lost their ability but more likely that they are having an off day, which happens to us all from time to time.

Despite this, the most wonderful information can be imparted from the realms beyond earth from some of the enlightened souls who reside there. If what they express can be accurately received by those who act as human antennae and then interpreted correctly, the most sublime wisdom emerges that inspires, uplifts and guides humanity along its path. This has always been the case and mediumship is not a new phenomenon, having been around for centuries in one form or another, even though in times gone by those that exhibited the gifts of insight and prophecy may well have been punished for their 'sins'. Today, mediumship seems to be flourishing with increased tolerance and more importantly greater mainstream interest being afforded this most beautiful of natural human abilities. All that is required now is for spiritualism to once more return to its heyday and embrace a more cutting-edge approach to working with the higher realms.

In the days of Arthur Findlay (1883-1964) - one of the great pioneers of spiritualism, and others, there seemed to be a more scientific approach and a genuine desire to push the boundaries of learning. My own branch of mediumship, involving the trance state, was commonplace and the quality of communication more reliable and accurate. Good physical mediumship was also in evidence, probably because the individuals capable of producing it had more time to develop properly without the myriad of distractions that most of us have around us today. What seems to be happening now though is that those who, having moved on to the next phase of life still desire to further advance and improve contact between the higher and lower realms are working with today's mediums and sensitives in new and exciting ways to prove the continuity of life. Spirit scientists continue to study energy and find new ways in which to adapt it with developments in the fields of EVP (Electronic Voice Phenomenon) and ITC (Instrumental Transcommunication) amongst others, pushing forward the boundaries. Work to enable moving images like those we stream through our computer laptop or TV are also in the pipeline and it cannot be too long before a major breakthrough happens allowing direct contact between dimensions without the need for the presence of a medium. Quite what some with terminally closed minds will make of this we can but guess?

The emerging realisation

It would be wrong of me to give the impression that only a select few such as mediums, savants and other sensitives form the basis of the paradigm shift. They may be at the cutting edge of change, but there are numerous other individuals and groups that are emerging worldwide to challenge the old belief systems and help break down barriers to understanding. As *(Fig 49.)* shows, there will always be those at the front of the queue – the trailblazers if you like, that lead the way for others to follow, but just as the crest of a wave needs the support of a large volume of moving water beneath it, so we find that this too is happening in the case of mankind. Despite the apparent 'dumbing down' of human behaviour and thinking reflected in some areas of society, there are by contrast many wonderful souls choosing to experience life as a human being at this time and this is not by accident.

As with any period of true and lasting change, the old systems have to crumble to make way for the new ones. This is difficult for them because they have no desire to surrender their grip on whatever portion of the human mind they appear to have control over. Religions don't want to give up their faith. Banks don't want to relinquish their grip over global finance. Governments can't bear the thought of letting go of their perceived power. Corporations don't want a downturn in profits threatening their livelihood. Energy companies don't want to consider alternative, clean, free energy. The stock markets fear a lack of 'growth' and the elite families and oligarchs who believe they have a divine right to rule over the masses, fear the loss of their 'wealth'. And so it goes on.

Without venturing too far into the realms of conspiracy theory, I have been aware for a considerable time that the human race is being manipulated by a relatively small number of people operating behind the scenes. These shadowy figures are unknown by the majority of people and their agenda, enforced by an ever expanding global system of control measures designed to manipulate the entire race for its own purposes has contributed greatly to the enslavement of many over time. But that is about to change as mankind awakens and *infinite consciousness* emerges once more to lead us toward a new Golden Age. We are starting to remember what we have almost forgotten and discovering the true essence of our being. We have come through the dark ages and are now emerging from the era of left brained intellect and reasoning which, although it served us well for a time, left us imbalanced and fragmented. From thinking and believing that we were separate, isolated individuals we have come to understand that we are all linked, part of one eternal whole. From believing that consciousness was a product of chemical and electrical processes occurring within the brain, we have come to know that *we are that consciousness* and that our brain is but a temporary vehicle for its expression. From being informed by science that at the end of life we are no more, we have come to understand from that same source what mediums have long since proven – our individual personality survives death. From believing that our earth was the centre of the universe we have come to know that it is but a tiny speck within the vast unfolding cosmos. From thinking that the world we see 'out there' is real and solid, we have come to understand that it is a holographic projection created from the waveform information that we decode from moment to moment. From believing that past, present and future were part of something called time, we have come to understand that this too is different than we had ever imagined and that space-time is itself a movement of consciousness within the eternal now. From believing on the one hand in a deity holding human form who sent his only offspring to save us, to asserting on the other that no creator exists, ever more us have come to understand that *we are the creator* and that consciousness is found wherever we look.

We have journeyed far, yet still play within the kindergarten of life. We have taken a billion steps forward in evolutionary terms yet struggle to stand erect as a spiritually awakened race. It seems that we have acquired much, but learned little. In true cosmic terms, we are probably considered by other more advanced intelligences to be a savage species - stupid, stubborn and ignorant in so many ways. When seen from the material perspective this is undoubtedly true. Yet viewed from the soul's perspective, it is anything but the truth. The universe is a sentient being and when considered as a complete entity, *is in the process of becoming – of knowing itself.* We are part of that unstoppable, majestic movement, essential fragments of the whole. Within us is all that there ever was, is and shall be. The rest is as they say, is but an illusion.

Summary of main points:

- We share relationships with life at every level
- A paradigm shift is underway part of which includes a unification of the scientific and the esoteric
- New precursors of evolutionary change are emerging
- Sensitive individuals act as harbingers of change and challenge the 'norms' within society as well as offering insight into the extraordinary ability we all possess
- The universe itself may be considered a sentient being and each one of us is an essential part of the whole

Epilogue: Epiphany

Sitting alone on the grassy knoll he watches the gentle glide of the white tailed eagle latching on to rising currents of warm air that take it to ever new heights in the azure blue sky. The freedom of the majestic bird mirrors his own as he breathes a deep sigh of contentment. His mind, still as sharp as it ever was reflects on the accomplishments of his life. At a hundred and fifty seven years of age, he knows that he can still teach his great, great grandchildren a thing or two. Below him the river glistens, reflected upon its surface, the rows of small but perfectly shaped domed houses, some with their glass roofs open to the warmth and light of the summer day. Each has its own source of power – a device utilising the torus dynamo to create free energy with absolutely no pollution. He allows himself a wry smile as he considers for a moment how the inventors of the device had at first been ridiculed and then persecuted by those who had thought the very concept of free energy absurd and wished to protect their own interests. He recalls too how wide-spread this activity had been back then.

His mind drifts as thoughts of other events that had shaped the course of human history filter into his consciousness. Mankind had been through so many trials and tribulations over the centuries, struggle upon struggle just to survive. Endless wars had been fought and millions had suffered and died at the hands of the ignorant and power thirsty few whose desire for control of the human race and the planet had been their ultimate goal. There was a time when it had seemed almost inevitable that every man, woman and child born upon earth would be compelled to live their lives enslaved by the system – microchipped at birth and forced to do the bidding of the burgeoning fascist state. With the drive to make all paper money obsolete and ensure that every aspect of human life was monitored, regulated and controlled through electronic surveillance, the distinct possibility of a global totalitarian society had seemed inevitable. Yet this too had passed. The paradigm shift that had begun at the end of the twentieth century resulting in the spiritual awakening of millions worldwide had been unstoppable. It was as if a slumbering giant had stirred and the descent into spiritual darkness that had been creeping up over many years had been stopped in its tracks by the sheer numbers of those whose eyes had been opened for the first time. The few had become the many and the many had morphed into one united force for good, galvanised into action from the level of their collective unconscious. All that had been hidden had been revealed as the vibrations of truth swept across the globe. It had been the worst of times and the best of times but now it was a truly wonderful time.

Scanning the horizon further he marvels at the new world that has unfolded over the last century. The utopian society lying before him is one of sheer beauty and refinement. Gone are the ugly skyscrapers dedicated to the servitude of the large corporations. No more are to be found the endless arrays of surveillance cameras and automated machines that had spied on every human activity. Gone are the many forms of transport and power stations that relied upon fossil fuels or poisonous nuclear energy. Free energy is abundant, powering everything from household devices to small cars and numerous saucer-like craft that can travel effortlessly beyond the earth's gravity, reaching out to other worlds. Gone too are the brutal military police and the controllers who carried out the bidding of their unseen masters. Now, life is sweeter, kinder and gentler, with the caring, nurturing spirit of an enlightened race prevailing.

Life expectancy has quadrupled and hospitals employ more holistic methods to treat patients. Prevention is the prime source of care with the emphasis on pure, natural foods. The genetically modified crops and artificial additives that wreaked so much havoc with the human body have been consigned to the past. Alternative healing methods have been integrated into mainstream medicine with an emphasis on light and sound as primary healing tools. The knowledge of words and language patterns to influence DNA has also evolved into an exact science and the power of thought and imagination to destroy tumours and repair tissue has largely replaced the surgeon's knife.

Children are taught to live as spiritual beings with more emphasis placed upon right-brained activity. Their creative skills are nurtured with music, literature and art being given prominence from an early age. Science has been merged with spirituality to offer a broader understanding of the true nature of existence and students are encouraged to express their unique individuality.

Although some money still exists enabling large transactions to take place, many smaller, basic interactions feature only the simple exchange of services by groups and individuals allowing everyone to offer something of value to the functioning of society with no-one excluded. People who do menial tasks are valued just as greatly as those who have governmental posts or positions that involve making important decisions and every individual is cared for by the state in times of need. Crime has also been reduced to manageable levels with severe sentencing reserved for only the worse type of offenders. Incredibly, the utopian society only dreamed of by some of the great philosophers of the past has become reality.

'How has all this been possible?' he ponders.

The answer forms in his mind almost as swiftly as the thought had arisen and his inner voice offers a simple yet succinctly profound reply.

'Through *unconditional love*' it whispers.

His eyes grow brighter and a smile crosses his lips as he contemplates the exquisite beauty of that short statement. 'Unconditional love' he thinks to himself, 'is the most powerful force in existence.'

'If only they had known that back then, they'd have saved themselves centuries of suffering.'

From a time when individuals had been encouraged to think and act only for themselves and where selfishness and egotism were considered virtues to which one should aspire, an astonishing and complete reversal had taken place. What had transpired over a relatively short space of time, perhaps fifty years at most, seemed almost inconceivable in the early part of the twenty first century. Yet without doubt a majestic transformation in thought and behaviour *had* occurred, driven by a deep and profound implicit order reaching forth from the very heart of life – a spiritual imperative that had all along been 'hard wired' into the very part of human DNA that scientists had considered 'junk'. Men had tried and failed to subjugate the human spirit and thwart the divine purpose of the creator and their lust for control and power had been the source of their own demise.

Although not a biblical scholar or religious in any way, he is reminded of a passage from Matthew 5:5 that his own grandfather had taught him as a child;

'Blessed are the meek for they shall inherit the earth'.

"It seems that even back then, all those centuries ago, someone knew what was to come".

Soon other thoughts begin to form in his mind accompanied by images, at first blurred and distant but then coming into clear focus as they draw ever closer. One in particular seems so real and vital like a distant memory from his past. He sees a figure; sitting huddled against a biting wind, scanning the night sky as if wondering what to make of the vast swathe of tiny pinpricks of light stretching across the heavens. He seems somehow familiar and yet……

Then another thought, vague at first, almost intangible yet growing ever stronger, bubbles up from the recesses of his subconscious mind and forms into a question that demands a reply…

"Is there something greater and more powerful than I?"

This time he knew the answer.

Appendix

Some core principles of the new paradigm

1) Physical 'reality' is a persistent illusion created by the mind.
2) Matter is not solid but appears that way because of the resistance between two electromagnetic waveform states.
3) We live in a 'virtual reality' holographic universe.
4) Our mind/body 'computer' decodes information from the waveform states thus allowing our real nature *(infinite consciousness)* to experience what we term 'the world'.
5) Our individual consciousness decodes vibrational information from the metaphysical universe into other forms of information:

   a) Electromagnetic
   b) Digital
   c) Holographic

6) Decoding 'reality' can be compared to logging on to the Internet. *We* observe it and interact with it *through* a computer. Our body/mind acts as a computer for our consciousness and provides for us the means to 'log on' to the waveform information construct. By doing this the five main human senses decode *vibrational* information into *electrical* information which is then transmitted to the brain and our entire genetic structure where it is decoded into *digital and holographic* information that we experience as 'physical reality'. *(See Fig 54. In this section).*
7) The human body acts like a lens that focuses our attention upon a specific small range of frequencies.

8) We are *individualised consciousness,* an aspect of *infinite consciousness*, operating through *mind.*

9) Everything has a degree of awareness.

10) Through the holographic nature of creation every part reflects the whole.

11) Energy cannot ever be destroyed - it only ever changes state. We *are* energy.

12) DNA acts as a human antenna that both transmits and receives information from the waveform construct.

13) Space, Time and Consciousness *(Space-time-consciousness)* are parts of one continuum.

14) We are connected to everything because we *are* everything.

15) Our true nature is all-knowing and all-powerful. Humanity is waking up and remembering itself *(the paradigm shift).*

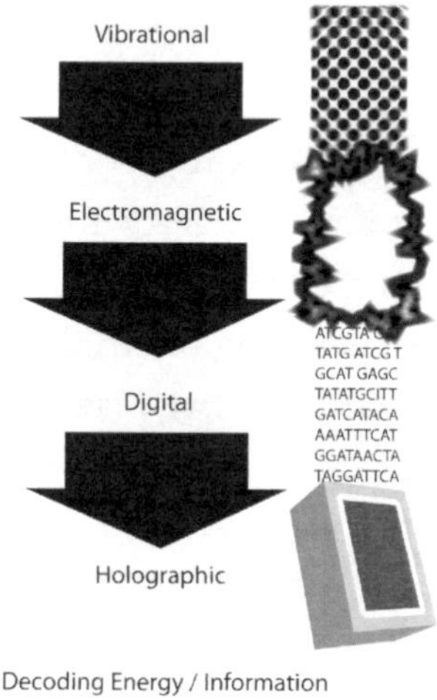

Fig 54. We decode energy (information) from the waveform construct

Brain frequency Beta - Highly alert and focused	14 -30 Hz
Brain frequency Alpha - Relaxed but alert	8 - 14Hx
Brain frequency Theta - Drowsiness (also first stage of sleep)	4 - 8Hz
Brain frequency Delta - Deep sleep	0.5 - 4Hz
Visionary Range	120 MHz
Bone	38-43 MHz
Brain frequency at 80-82 MHz	indicates a genius
Healthy body (neck down)	62-68 Hz
Thyroid and Parathyroid glands	62-68 Hz
Thymus Gland	65-68 Hz
Heart	67-70 Hz
Human cells start to mutate when their frequency drops below	62 Hz
Lungs	58-65 Hz
Liver	55-60 Hz
Pancreas	60-80 Hz
Stomach	58-65 Hz
If the frequency drops just 4 points this is when a headache will start	58 Hz
Disease begins, Like the cold symptoms	58 Hz
Flu invades the body	57 Hz
Viral Infection	55 Hz
when more serious problems come about like pneumonia, Epstein Barr and etc.	52 Hz
Tissue breakdown from disease	48 Hz
Cancer can set in	42 Hz
Death begins at	20 Hz

Fig 55. Human Bio-frequency table courtesy of Bruce Tainio of Tainio Technology in Cheney, Washington

Postscript

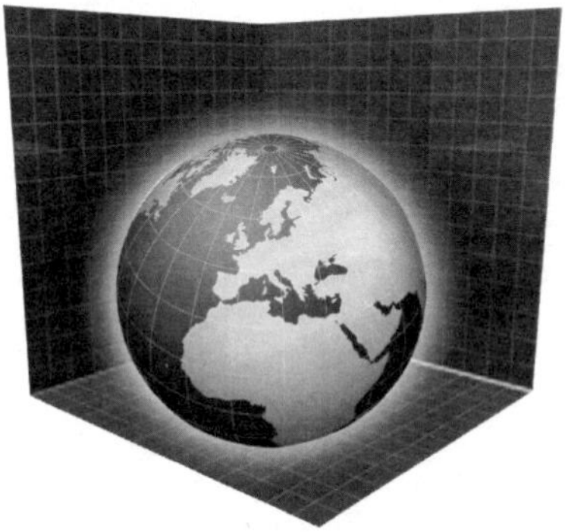

At the time of going to press, new studies undertaken by researchers at the University of Bonn in Germany suggest that the universe may be a 'computer simulation'. As incredible as this might at first appear, the evidence that is starting to emerge is intriguing if not yet compelling. The problem that arises with all simulations is that the laws of physics, which appear continuous, have to be superimposed onto a discrete three-dimensional lattice that advances in steps of time. This, by its very nature imposes limitations upon the fundamental operation of energised particles because nothing can exist that is smaller than the lattice itself. So if the universe *is* merely a simulation, there ought to be a cut off in the spectrum of high-energy particles – *which is exactly what has been discovered.* Silas Beane and his colleagues Zohreh Davoudi and Martin J. Savage (in a revised paper published on 9th November 2012) calculated that this seemingly arbitrary cliff in the spectrum is consistent with the kind of boundary that would exist if there was an underlying lattice governing the limits of a simulator.

In normal circumstances high energy particles interact with the cosmic microwave background resulting in a loss of energy over long distances and Beane states that "The most striking feature is that the angular distribution of the highest energy components would exhibit cubic symmetry in the rest frame of the lattice, deviating significantly from isotropy."

What this means is that if a lattice really does exist, the cosmic rays would prefer to travel along the axes ensuring that we would not see them equally in all directions. Current technology can simulate this effect and witnessing this would be equivalent to being able to 'see' the orientation of the lattice on which our universe may be simulated.

Naturally a great deal more investigative research needs to be done but if this does turn out to be consistent with a simulator lattice, then this points to several scenarios. It could show us that there's a boundary out there consistent with Beane's hypothesis, operating in ways that we'd expect if we were living inside a simulator (based on the same parameters that we would use) or it could be that we're incorrectly interpreting evidence of certain fundamental laws that we are as yet unfamiliar with.

Before jumping to any conclusions about whether the human race or the cosmos really is a simulation and by what or whom this was created, I offer the suggestion that this latest ground breaking research may point towards further evidence of the holographic nature of reality outlined in this book. Could the 'arbitrary cliff' or boundary be evidence of Bentov's 'skin' or 'surface' of the torus – or the point in his model at which the flow of matter begins to curve back upon itself? Could a 'lattice grid' be the cosmic version of ley lines? These concepts and others seem to me to be well within the bounds of possibility but until more evidence emerges I can only speculate on the wider implications of the findings and I remain open to all new thinking and discovery.

Robert Goodwin - *February 2013*

The published paper can be sourced here:
http://arxiv.org/abs/1210.1847

Bibliography

1. **Talbot,** M., *The Holographic Universe:* Harper-Collins, 1991

2. **Icke,** D., *Remember Who You Are:* David Icke Books, 2012

3. **Bentov,** I., *Stalking The Wild Pendulum:* Destiny Books, 1988

4. **Kelly,** Dr. R., *The Human Antenna:* Energy Psychology Press, 2007

5. **Lipton,** B., *Insight Into Cellular Consciousness:* Bridges,2001, Vol 12(1):5

6. **Wolff,** M., *Tetrode and EPR: Matter Waves and Buddhist Thought*

7. **Kamenetskii,** F., *Unraveling DNA:* Perseus Books, 1997

8. **Lipton,** B., *The Biology of Belief:* Mountain of Love/Elite Books 2001

9. **Rein,** G., *The Body Quantum: Non-classical Behavior of Biological Systems,* "The Resonance in Residence Science Addendum," Ilonka Harezi, 2002

10. **Oschman,** J., *Energy Medicine; the Scientific Basis:* Churchill Livingstone, 2000

11.**Wordsworth,** C.F., *Holographic Repatterning Modalities for Transforming Resonance Patterns:* Wordsworth Productions, 1994

12. **Linsteadt,** S., *The heart of health: the Principles of Physical Health and Vitality,* 2003

13. **Braden,** G., *The Divine Matrix:* Hay House UK, 2007

14. **Sheldrake,** R., *The Nature Of Formative Causation:* Park Street Press, 2009

15. **Wells,** J., *The Politically Incorrect Guide to Darwinism And Intelligent Design:* Regnery Publishing, 2006

16. **Bryant,** P., *Starwalking:Shamanic Practices for Travelling into the Night Sky:* Bear & Company, 1997

17. **Ring,** Dr. K & **Cooper**, S., *Mindsight: Near and Out of Body Experiences in the Blind:* Institute of Transpersonal Psychology, 1999

18. **Newman,** F., *The Universe of Silver Birch:* Psychic Press, 1994

19. **Bischof,** M., Biophotons – *The Light in Our Cells:* Zweitausendeins, 1995

20. **Pasricha,** S., *Journal of Scientific Exploration* Vol 15, No 2, pp 211-221: 2001

21. **Newton,** Dr. M., *Life Between Lives - Hypnotherapy*: Llewellyn Publications, 2004

22. **Leininger,** B, A., *Soul Survivor:* Grand Central Publishing, 2010

23. **Borgia,** A., *Life in the World Unseen:* Odhams Publishing, 1954

24. **Edwards,** H., *Life in Spirit:* Healer Publishing, 1976

25. **Eagleman,** D., Incognito: *The Secret Lives Of The Brain:* Canongate Books Ltd, 2012

26. **Tolle,** E., *The Power Of Now:* New World Library, 1999

27. **Hawking,** S., *A Brief History of Time:* Bantam Books, 1998

28. **Aurobindo,** S., *The Life Divine:* Sri Aurobindo Ashram Publications Department, 2001

29. **McTaggart,** L., *The Field:* Harper Collins, 2001

30. **Treffert,** D., *Extraordinary People: Understanding Savant Syndrome:* Harper Collins, 1989

31. **Treffert,** D. *Islands of Genius:* Jessica Kingsley Publishers, 2010

32. **Aron,** E.N., *The Highly Sensitive Person: How to survive and thrive when the world overwhelms you:* Thorsons, 1999

33. **Aron,** E.N., *The Highly Sensitive Child: Helping our children thrive when the world overwhelms them:* Thorsons, 2003

34. **Sagan,** C., *The demon haunted world: Science as a candle in the dark:* Ballantine Books, 1997

35. **Alexander,** E., *Proof of Heaven:* Piaktus, 2012

Internet Sources

http://humandna.co./in

http://www.rense.com/general62/expl.htm

http://www.fosar-bludorf.com/index_eng.htm

http://www.abc.net.au/science/articles/2001/04/04/133634.htm

http://library.thinkquest.org/22601/advanced/cope.html

http://www.peterrussell.com/SG/ch4.php

http://med.stanford.edu/ism/2012/may/endy.html

http://www.integralscience.org/sacredscience/SS_quantum.html

http://aminotes.tumblr.com/post/7722763662/david-eagleman

www.bluetreeyoga.com/pdf/WaterHasFeelingsToo.pdf

http://suite101.com/article/resonance-and-the-human-antenna-a377373

http://www.naturalhealinghouse.com/articles_scalar-waves.php

http://pesn.com/2011/03/26/9501797_Teslas_Scalar_Waves_Replicated_by_Steve_Jackson/

http://www.pelorian.com/multiverse/dna.html

http://www.zengardner.com/change-the-morphic-field-change-the-world/

http://radyananda.wordpress.com/2012/02/27/911-foreknowledge-caught-on-computers-that-measure-global-mind/

http://www.brainyquote.com

http://blog.gaiam.com/quotes/topics/quantum-reality

http://www.angelfire.com/space2/light11/nmh/arrival1.html#nature

http://www.energyreality.com

http://www.awakening-healing.com/Healing/principles_of_energy_medicine.htm

http://www.near-death.com/evidence.html#a3

http://www.fullspectrum.org.uk/about-kirlian/info_13.html

http://transpersonal.de/mbischof/englisch/webbookeng.htm

http://www.geo.org/dowse1.htm

http://www.reversespins.com/famousquotes.html

http://www.scientificexploration.org/

http://www.spiritualregression.org/page.php?slug=life-between-lives

http://www.reversespins.com/proofofreincarnation.html

http://www.thenewearth.org/LifeAfterDeath.html

Internet Sources

http://thinkexist.com/quotes/ralph_waldo_trine/

http://www.cracked.com/article_19659_7-theories-time-that-would-make-doc-browns-head-explode.html

http://www.nist.gov/public_affairs/releases/aluminum-atomic-clock_092310.cfm

http://library.thinkquest.org/06aug/02088/hawking.htm

http://einstein.stanford.edu/SPACETIME/spacetime2.html

http://auromere.wordpress.com/cosmology/spacetime/

http://sriaurobindostudies.wordpress.com/2010/05/23/space-time-and-consciousness-part-3/

http://www.sriaurobindoashram.org/ashram/mother/index.php

http://www.goodreads.com/quotes/tag/universe

http://spider.ipac.caltech.edu/staff/jarrett/2mass/LSS/

http://www.big-bang-theory.com/

http://www.lynnemctaggart.com/blog/165-making-something-out-of-nothing

http://odewire.com/48044/the-amazing-promises-of-the-zero-point-field.html

http://www.ecolo.org/lovelock/what_is_Gaia.html

http://www.heartmath.org/research/science-of-the-heart/introduction.html

http://www.professional-house-clearing.com/curry-grid.html

http://www.biogeometry.org/page34.html

http://www.bmedreport.com/archives/7303

http://www.earthbreathing.co.uk/sr.htm

http://www.addictiontherapyclinic.com/frequency.html

http://www.esotericscience.org/

http://www.neatorama.com/2008/09/05/10-most-fascinating-savants-in-the-world/

http://www.worldscientific.com/doi/abs/10.1142/S0219635203000287

http://ritacarter.co.uk/page18.htm

http://www.hsperson.com/index.html

http://www.eternea.org/Eben_Alexander/biography_eben.htm

http://arxiv.org/abs/1210.1847

http://www.technologyreview.com/view/429561/the-measurement-that-would-reveal-the-universe-as-a-computer-simulation/

http://www.newscientist.com/article/mg21628950.300-the-idea-we-live-in-a-simulation-isnt-science-fiction.html

About the author

Robert Goodwin was born in England in 1954. He grew up in his native city of Birmingham and pursued an interest in mediumship during his late teens when he became aware of the latent ability he had to connect to the afterlife. After sitting in a small group for several years to develop his gift, he began demonstrating publicly in 1979, something he continues to the present day.

Although a fully qualified Hypnotherapist and NLP practitioner, Robert is best known for his work as a trance medium and along with his wife Amanda, frequently demonstrates throughout the UK and Europe. Together they have published several books of spiritual philosophy and their work has also been featured in Psychic News, the UK's foremost Spiritualist publication. Robert has also appeared on local radio and his popular website features recordings of his interviews and clips of his trance work.

Robert also gives private readings from his home in Kidderminster and runs development groups for those wishing to enhance their own mediumship.

www.whitefeather.org.uk